Build Your Own Printed Circuit Board

Al Williams

McGraw-Hill

New York   Chicago   San Francisco   Lisbon
London   Madrid   Mexico City   Milan   New Delhi
San Juan   Seoul   Singapore   Sydney   Toronto

The McGraw-Hill Companies

Cataloging-in-Publication Data is on file with the Library of Congress

1 2 3 4 5 6 7 8 9 0   DOC/DOC   0 9 8 7 6 5 4 3

P/N 142830-5
PART OF
ISBN 0-07-142783-X

*The sponsoring editor for this book was Judy Bass and the production supervisor
was Sherri Souffrance. It was set in Century Schoolbook by MacAllister Publishing
Services, LLC. The art director for the cover was Margaret Webster-Shapiro.*

Printed and bound by RR Donnelley.

McGraw-Hill books are available at special quantity discounts to use as premiums and
sales promotions, or for use in corporate training programs. For more information,
please write to the Director of Special Sales, McGraw-Hill Professional, Two Penn
Plaza, New York, NY 10121-2298. Or contact your local bookstore.

For my wife, Pat, with all of my love.

CONTENTS

Contents

Contents

ACKNOWLEDGMENTS

While I was writing this book, the United States, Great Britain, and a few other countries invaded Iraq. Regardless of what you thought about the war, politically and morally, you could not deny that the video sent by satellite by embedded reporters was astounding. Watching live footage of warfare gave most of us the closest taste we will ever have (hopefully) of the sheer terror the young men and women on both sides face.

What I found most interesting, though, is that the front line of an army is like the point of a knife. Behind every solider there is an enormous support apparatus that isn't apparent from the news video. Quartermasters and other solders provide the front-line solder with food, clothing, ammunition, fuel, and intelligence among other things. Even the payroll clerk that handles the soldier's pay is vitally important. But you rarely, if ever, see any of these people who make the front line job possible on the news channels.

Books are much the same way. My name may be on the cover, but in fact, I'm just the tip of knife. The book you hold in your hand is possible only because of a tremendous infrastructure that includes editors, typesetters, printers, publishers, and marketing staff. This team—including Judy Bass and Scott Grillo at McGraw-Hill and Beth Brown and her team at MacAllister Publishing Services—made this book, as you see it, possible and they have my gratitude for it.

I'd also like to thank the folks at CadSoft (especially Klaus Schmidinger) for creating Eagle and allowing us to distribute it with the book.

My family—which includes my wife Pat, three kids, three grandkids, two dogs, and a cat—also have my thanks. Not only did they put up with another book, but they also didn't complain too much about the ferric chloride stains on the patio!

INTRODUCTION

Electronics is a paradox. It has become simpler and more complex at the same time. In the 1950s and '60s you could easily experiment with electronics with just a few simple hand tools. However, most people could only design and build relatively simple projects. Today, you have a bewildering array of sophisticated microprocessors, analog functions, and programmable logic arrays. A small company or even a hobbyist can produce his or her own real-time controllers, video signal processors, and many other circuits that would have taxed a huge research laboratory back in the '60s.

However, these large-scale projects come with a price. Although these projects may be easier than ever to design, they are becoming more difficult to build. An increasing number of components are available only in surface-mount technologies. Even for through-hole components, higher operating frequencies and component density makes traditional prototyping methods difficult or impossible to apply.

The good news? It is now easier than ever to design and build *printed circuit boards* (PCBs or PC boards) in a small lab setting. There are several technologies—notably personal computers, the Internet, and laser printers —that have combined to make PC board production much easier than ever before.

- Inexpensive (even free) software for PCB design. (This book contains a free copy of the powerful Eagle software, for example.)
- Using a laser printer, it is possible to transfer a finished PCB design to a copper board. Although you'll still need a few chemicals, you **won't** need a darkroom or much in the way of special equipment. The process is **not** photographic which makes it much easier to make small numbers of boards.
- If you want a high-quality board produced, the Internet makes it easy to get a quote and order a board from anywhere in the world.

Before inexpensive *computer-aided drafting* (CAD) and *computer-aided manufacturing* (CAM) tools, it was a major undertaking to send your artwork to a board maker who wasn't nearby. If you wanted to take advantage of production in another country (which may be substantially less expensive than a domestic board maker), there was the added problem of paying for the boards and shipping the finished product to you. Between the Internet, modern credit card payment, and excellent mail and express delivery companies, importing boards is now far easier than ever.

Is This Book for You?

If you are designing or building any type of electronic circuitry, you can benefit from PC boards. Projects built with PC boards look more professional than wire wrap or hand-wired boards. In addition, modern PC boards tend to have fewer wiring errors (CAD programs check for obvious errors) and work better at high frequencies. If you are working with surface-mount components, a PC board of some kind may be the only way you can build a prototype.

If you are a design engineer, a hobbyist, or a student, you can use the information and software in this book to design high-quality PC boards and either make them yourself, or have them professionally made at a reasonable cost.

Before you start this book, you should have a basic understanding of electronics. This book won't explain how to read a schematic, for example, or how to identify a resistor. You should also have basic familiarity with computers. If you can operate a word processor or a Web browser, you should be fine. You'll also need a computer to actually run the software. Eagle will work with most Windows or Linux computers. If you want to physically produce your own PC boards, you'll need some other equipment, but you can wait until you get to those chapters to decide which method you want to use (and therefore, what equipment you'll need).

If you are currently building prototypes on perf board or using wire wrap techniques, you'll be surprised at how easy it is to produce excellent PC boards with only modest tools. If two designers were trying to sell you a new idea, which would be more credible: one with a circuit on a professional-looking PC board, or one that had a rat's nest of wire and solder? There is no reason to not have that professional look with all your prototypes and projects.

If you are a professional designer and someone else lays out your final boards, you'll still find it useful to have a better understanding of modern PC board layout techniques. Having a clear idea about the board layout (and maybe even a first draft) will make the final layout smoother, and you'll be more satisfied with the results.

Getting Started

If you are ready to start making your own boards, you'll want to begin with —logically enough—Chapter 1, "Printed Circuit Fundamentals," and follow through to Chapter 7, "Eagle Output" (you may want to just skim Chapters 5 and 6 if you are in a hurry and don't need custom libraries or scripting).

After you've read the core chapters, you may want to skip around depending on your needs. For example, Chapter 8, "Boards from a Laser Printer," discusses how to easily make simple boards using a laser printer and a few simple chemicals. Chapter 9, "Photographic Boards," covers using the traditional photographic method to produce boards. This is a lot more work but has certain advantages over the laser printer method. If you want to send your boards to a professional board house, you'll want to skip directly to Chapter 10, "Outsourcing Boards."

Regardless of the methods you choose, you won't design any boards reading this introduction. There is a Chinese proverb: "The longest journey begins with the first step." Although the journey to PC boards isn't really that long, the first step—Chapter 1—is on the next page. The sooner you begin, the sooner you'll be admiring your first PC board.

CHAPTER

Printed
Circuit
Fundamentals

✳ In 1798, Eli Whitney secured a government contract to produce 10,000 muskets for $130,000. Many didn't believe he could do it because producing guns at that time required skilled gunsmiths. Also, because each gun was slightly different, parts were not interchangeable. What Whitney envisioned, and successfully implemented, was a system where precise machine tools would produce individual parts of exact sizes and shapes. Operated by relatively unskilled labor, these machines would produce parts that could be assembled into many identical rifles. Other industrialists—Henry Ford, Cyrus McCormack, and Isaac Singer (of sewing machine fame)—were quick to follow suit, and the result was the Industrial Revolution.

The field of electronics has had several similar revolutions. If you looked into an old-fashioned radio receiver, the first things you'd notice would probably be the tubes. But, after the tubes, you'd probably notice the mess of wires connecting all the components together. Someone had to physically connect those wires and solder them. Of course, that means mistakes happen—you can't expect people to build radios without making errors. The more complicated the project, the more mistakes you can expect.

In the 1950s, companies started using *printed circuit boards* (PCBs) to reduce costs and errors. The idea is quite simple. You start with an insulating substrate (the board) made of phenolic (a material made from paper and resin), fiberglass, or ceramic. Then you arrange for copper to be placed on the board in a particular pattern. This copper, which is bound to the substrate, conducts the electricity just like wires. Holes are drilled to allow components to mount to the board.

Offhand, this doesn't sound much better than wires. The trick is that automated processes can create the copper patterns on the board easily. Just as Whitney's machines turned out perfect rifle barrels, it is possible for modern manufactures to mass-produce perfect circuit boards. Of course, the trick is how to actually do this. Several possible methods are available. You ✳ can take a board that already has copper laminated to it and remove the copper you don't want. This is usually done with a photographic process and chemicals that attack copper. However, it can also be done with a computer-operated milling machine that mechanically chips away the copper you don't want.

Other ways to build PCBs exist. For example, you can start with a blank substrate board and electroplate the copper on the board. Some PCBs are not made photographically, but are made with a silk screen.

No matter what the method, the key is that, armed with some negatives (or a milling machine program), you can produce dozens or hundreds or even tens of thousands of boards, all precisely the same. The cost of producing these boards is relatively inexpensive, and there can be other advantages as well. For example, with precisely placed components, it is possible to use machines to insert the components into the board. Another machine can solder the parts, practically eliminating human involvement in the production process.

When PCBs first became widespread, producing them was high technology indeed. First, artwork (the pattern that shows where the copper should be on the board) was done by hand. Because the tolerances were important, designers typically used drawings that were four times larger than life size. Then a special camera would take a picture of the artwork and reduce it. The resulting film would then be used in a complex manufacturing process.

Several things have happened to make PCBs more accessible than ever to the small company or hobbyist. First, computer programs are available that enable you to design PCB layouts easily. As with anything you do on the computer, you can correct errors, experiment with different layouts, and make changes very easily. You can even have the computer do a great deal of the work for you.

The industry has a standardized format for describing PCB layouts. When you are satisfied with your design, you can simply e-mail your files to any number of companies that will build the board to your specifications and ship it back to you. If you are willing to pay enough, you can design a board on Monday and have it in your hands by Wednesday.

However, it is also possible to build your own boards with a modest investment in tools and chemicals. You can create film with a common computer printer. You can also use a laser printer (or copier) to directly transfer the pattern to the PCB material. Then several inexpensive and relatively safe chemical processes can be used to finish the job.

Like the gun, what was once the province of specialized labor is now accessible to anyone who wants to build electronic circuits. That's a good thing, too. Modern circuits are getting harder and harder to build by hand. Working with normal integrated circuits is hard enough. With the trend toward tiny surface-mounted parts, PCBs are almost a necessity, even for prototyping or one-off projects.

A PCB

Figure 1-1 shows a typical PCB. Like most boards these days, it has multiple layers of copper (this board has two layers). A homemade board is easier to make with a single layer, but when having boards made commercially, two layers are practically as inexpensive as a single layer. (Because you can

almost always make a two-layer board smaller than one with a single layer, two-layer boards are typically less expensive than a single-layer board to produce.)

In fact, designing a board like one in the photograph requires more than just two layers. This board has six different design layers:

- The bottom copper layer
- The top copper layer
- The markings on the top side that identify the components
- A mask layer for the bottom
- A mask layer for the top
- A layer that defines where the holes are drilled

The mask layers are not visible, but they define where a special chemical coating prevents solder adhesion to the board. This makes it much easier to build the board by hand and is essential for automated soldering equipment.

Even the innocent-looking holes in the board are special. First, they must be precisely placed so that the components will fit. Integrated circuits are especially finicky about hole alignment. Also, on a two-layer board, the holes are plated internally with copper so that the top and bottom of the board are electrically connected at every hole. Most of the holes are for fitting components, but some exist just to connect a copper line (known as a track or a trace) on the top to one on the bottom. These special holes are known as *vias*. Plated through holes also increase the board's strength because it is harder to delaminate the copper from the substrate when the copper is attached on both sides.

Because this is a professionally produced board, it also has a tin or solder plating on top of all the copper. This makes the copper easier to solder and protects the copper against corrosion.

Homemade boards, like the one in Figure 1-2, tend to be simpler. This board has a single copper layer, no component markings or masks, and no plating. Because it was drilled by hand, the holes are not as precisely placed as the other board.

Typical Design Flow

If you don't think too hard about it, it seems that laying out a PC board would be easy. The problem is that you can't simply put components anywhere you want on the board. Of course, electrical issues must be considered. For example, a crystal or bypass capacitor might need to be close to its associated integrated circuit. However, the most vexing problem with layout is that copper traces on the same side of the board must not cross (unless they are supposed to connect, of course). Unlike wires, a copper track cannot be laid on top of another—except by placing it in another layer.

Figure 1-2
A homemade PC board covered with a coating to protect the copper surface

It is somewhat more difficult, but you can in fact lay out boards with more than two layers. For most small uses, this is too difficult to fabricate, build, and repair, so unless you are working for a large company, you probably won't run into many boards with more than two layers. Most PC motherboards, for example, have multilayer boards.

Nearly every PCB begins life as a schematic. Schematics don't directly relate to the PCB—they are just an abstract representation of the circuit you want to build. However, the schematic does contain all the connections the circuit requires to operate properly. It is possible for the designer to simply lay out the board by looking at the schematic, but modern tools give designers a better method: schematic capture.

Schematic capture is a fancy way of saying "drawing a schematic on the computer." Sometimes you'll copy a schematic from paper or another program into your *Computer-Aided Design* (CAD) program. If you are the original circuit designer, you probably just drew the schematic in your CAD package in the beginning. Even if you are using a different CAD package than the designer, many computer programs can export their schematic or at least a net list (a computer file that describes a schematic in a machine-readable format).

Once you have the schematic, you can start the layout. You'll need to decide on the size of the board (which is often dictated by other demands). You arrange the components (or at least make an attempt at arranging them) and then have to define the traces that connect the components together so they match the schematic.

You can do this manually—it is reminiscent of solving a maze or a puzzle. However, many CAD tools include an autorouter that will try to fit the traces so they match the schematic. Some of these do a better job than others, and it is a rare board that doesn't require a little manual intervention to successfully route.

This is often an iterative process. After you find yourself at an impasse, you may have to rearrange the components on the board and try again. If you are really desperate, you might even make the board larger to give yourself more room.

Once you have the board laid out, you can tweak the additional items, such as the component marking. Hopefully, you'll have room for mounting holes and any other extras you need to include.

At this point, you'll usually print out a copy of the artwork and examine it closely to make sure everything will really fit. Most CAD programs can also check to make sure you don't have any obvious blunders (such as wires not connected, traces touching, and so on).

Your next step depends on how you'll produce the board. If you are going the professional route, you'll generate some Gerber files (machine-readable files that describe the layers on your board) to describe the board for the automated machines that will produce the board. Files that describe the drills are usually in Excellon format (which is just another machine-readable file format). You can also use these files to have a photographic image professionally made (or a photo plotted). You can use it to make your own board.

If you are going to produce your own board, you'll probably just make a high-quality 1:1 printout of the artwork onto film or certain types of paper. You'll use the artwork to make a layer of resist on a piece of special copper-clad board and then use a special chemical that removes the copper from any place that isn't covered with resist. When you remove the resist, the copper that is left is your circuit design.

Making your own board also means drilling holes. An ordinary drill is not suitable for this job because the tiny drill bits you need break with any horizontal movement. A regular drill press may work, but it isn't really made for this purpose. Not only is it usually too slow, but it has *runout*, which means the bit makes a small circle instead of spinning exactly on its center line. A better option is a Dremel tool in a drill press attachment or a jeweler's press (which is a special drill press made for fine work).

Either way, you wind up with a finished board, ready to build. This sounds fairly simple, but as you'd expect, the details are where you get stuck. You'll explore each of these steps in future chapters. To summarize, the basic flow is as follows:

- Schematic capture
- Board layout
- Routing
- *Computer Aided Manufacturing* (CAM) output or film creation
- Board production

Don't confuse routing—the design of the copper pattern on the board—with the machining process known as routing. Usually, boards are mass-produced by making a single large board that has several repeated patterns. When completed, the boards are cut apart to produce individual boards. This is also known as routing (if it done with a router). In some cases, the boards are separated with a sharp blade (known as shearing). Routing is preferred if the board is not square or rectangular or has slots cut in the interior of the board.

Other PCB Advantages

You've already seen that PCBs can reduce wiring errors, decrease assembly costs, and improve the prospects for an automated assembly. PCBs have other advantages as well. Because mass-produced PCBs are uniform, it is possible to take advantage of their physical properties. For example, you might design a long trace of copper to form a small current-sensing resistor. In radio work, making small-value capacitors and inductors from PCB patterns is quite common. Many wireless devices (such as *Wireless Fidelity* [WiFi] network cards) use board patterns for antennas. Solder masks—a coating that repels soder from areas where it does not belong—are another advantage of printed circuitry. A board with a solder mask is easy to assemble. For high-density surface-mount parts, a solder mask is almost essential to prevent solder bridges between pins. Unfortunately, solder masking is not something you can usually do yourself, so if you require this feature, you'll have to plan on having your boards made professionally.

PCBs typically consume less space than a traditionally built circuit. This is particularly true when using surface-mounted components. Cell phones, personal organizers, and laptop computers could not be as small as they are without surface-mounted components and printed circuit boards.

Are there any disadvantages to PCBs? Yes. First, it takes a small investment to produce PCBs (although with the free software included in this book, you are ready to produce boards that will be made by a professional shop). Also, although PCBs reduce wiring errors in a repetitive assembly, if you make a mistake in the layout, it can be difficult and expensive to fix.

The Economics of PCB Production

The next time you are at an office supply store, ask them how much they'd charge you to print five business cards. They will probably tell you they won't do it. Then compare the price between 100 cards and 500 cards. All the cost in printing is up front. The primary expense is setting up to print that first card. Once you print the first card, the second card adds just a few pennies to the cost.

PCBs are the same way. The professional production of a board requires _tooling_—the production of reusable tools used in the manufacturing process. Most professional board houses will charge you $100 to $200 to produce these tools. That's $200 and you haven't even bought one board! However, you only pay this fee once (a *nonrecurring engineering* [NRE] *fee*). The actual cost per board depends on several factors, including the board's size, the materials used, the number of holes drilled, and other factors, such as how long you are willing to wait for the boards.

Some specialty companies produce prototype boards and may not charge an explicit NRE fee. If you are willing to have work done outside the United States (which means a longer shipping delay), you can get large boards made for $20 or $30 each.

The biggest drawback to an NRE is that any change to your board, no matter how slight, will require you to pay the NRE again. If you are on good terms with your board producer and you only need to fix one layer, they may (or may not) charge you just for redoing one layer. It pays to not make any mistakes when you use an outside board house. Some board manufacturers —eager to get your business—may waive the NRE fee, especially if they think you will order in large quantities.

Of course, if you produce your own boards, you won't have these expenses. If you make an error, you've wasted some time and ruined some materials (which are not that expensive). Still, you want to produce a workable board the first time, every time.

In truth, of all the boards I've ever produced, only a handful were absolutely perfect the first time around. However, you hope to produce a board that is usable and then fix any little problems on the next revision. For example, I'll often realize that there should have been more markings for components or that two parts are a little closer together than I'd like. However, you can usually use the boards (the worst case being that you might need to cut a trace with a hobby knife or add some extra wires) and justify your NRE investment until you are ready to make more boards.

Resources

You'll find a listing of several PC board makers in Appendix A. Some specialize in small jobs and others are mainly for larger production runs. Of course, if you plan to make your own board, these aren't necessary.

However, whether you make your own boards or have them sent out, you still need a CAD program. This book covers an excellent design program that runs under Windows or Linux called Eagle. Many other programs are available, but I like Eagle for several reasons:

■ A free version is available that is very usable. (The paid version can make larger boards with many layers, but the free version is more than adequate to design two-layer boards that are fairly large.)

- The free version includes schematic captures, board layout, and autorouting.
- The commercial version is relatively inexpensive.
- A substantial library of components is included.
- You can create your own libraries if necessary.

Other programs have been made for producing PCBs and you'll read about some of them in Chapter 12, "More Choices." However, Eagle is a great tool and free to try.

One thing you probably don't want to do is use a general-purpose CAD program to lay out your board. Sure, it is possible to lay out a board using AutoCAD or TurboCAD. However, these tools are not as good as a program that understands the relationship among a schematic, electrical signals, and the board. Some add-on modules turn programs like AutoCAD into a PCB design environment. However, an unadulterated general-purpose CAD program is usually a poor choice for laying out boards.

Of course, you could do worse. I've actually seen people lay out boards using programs like CorelDRAW or Visio (which are just ordinary graphic programs). Granted, in 1980, that would have been a boon compared to doing everything by hand. Today too many good tools (many of them free) exist to resort to a makeshift solution. Programs that deal with pixels (such as Microsoft Paint, for example) are completely out of the question. Because they deal with pixels, they won't output with enough resolution to be useful, but even programs that work with vector graphics (such as Corel) are not adapted for PCB layouts and you should avoid them.

One key feature you must have in any CAD program is a way to save the necessary files if the boards are to be professionally made. Even if you don't plan on having boards sent out, you never know when you might want to have boards made, and it would be a shame to have your layout locked into some odd format no one will accept. Some board houses will accept unusual file formats such as PostScript, but Gerber files are the standard. Of course, you can produce film, but that's difficult, plus it is becoming difficult to find manufacturers that will still accept film. On top of that, you have to physically send film, which delays the process. You can e-mail a Gerber file and the manufacturer starts working on your board right away.

A few companies offer you "free" design software. The catch is that the software doesn't produce Gerber or other industry-standard files. To produce a board, you must use their service. Some of these services are good, but they essentially have your design as a hostage. Some of them will produce Gerber files for a hefty fee, and others will simply insist on building your boards. If you want to send them somewhere else, you'll need to redesign your board using another software program. With so many programs that can produce industry-standard files, and so many companies eager to build your boards, forcing yourself to use these proprietary programs is unnecessary.

Materials Needed

If you are planning to send your boards somewhere for production, you don't need anything more than the CD-ROM in the back of this book and a computer to run the software (either a Windows or a Linux computer). However, if you want to make your own boards, you'll need a few other items:

- Etching chemicals
- Plastic trays or containers for chemicals (these can be common Rubbermaid or similar containers that seal closed)
- Blank copper boards
- A printer for your computer (preferably a laser printer)
- Basic safety gear such as eye protection, latex gloves, and a filter mask
- A high-speed rotary tool (such as a Dremel) with a drill press attachment, a jeweler's press, or—if you have nothing else—an ordinary drill press
- A jigsaw or hacksaw
- A file and sandpaper
- A selection of small drill bits (these are special bits you can't find at ordinary hardware stores)

Depending on the process you decide to use, you may need a few specialized items.

Safety and the Small Lab

If you intend to send your boards to be manufactured, then you don't have any special safety issues to worry about (other than, perhaps, sitting in front of a computer monitor). However, if you are planning on making your own boards, you need to be careful. None of the chemicals are especially dangerous, but you should still be cautious when working with any chemicals.

The most common chemicals used are ammonium persulphate (which is pretty innocent) and ferric chloride. Ferric chloride isn't very harmful per se, but it does stain everything it touches. You should also be familiar with any local laws that concern the disposal of chemicals. You can obtain *material safety data sheets* (MSDS) for these two chemicals from the company that sells them. A collection of common data sheets can be found at www.mgchemicals.zcom/msds/index.html.

Chemicals used in solder masks and tin plating may be less safe. Most of these processes are out of reach for most builders. Nearly all boards built by hand will not have solder masks. Several "electroless tin" solutions exist that aren't technically electroplating but coat your boards with a layer of tin. Liquid Tin (from MG Chemicals) works well at room temperature, but it is a potent acid. Other products (like TinIt) are not as handy because they require heating and care in handling.

Cutting and drilling fiberglass can also present hazards. In particular, some people fear that fiberglass dust may be hazardous to inhale (although the International Agency for Research on Cancer classifies it as "probably non-carcinogenic"). However, the dust can cause eye and skin irritation.

Drilling not only produces dust, but it can also produce a hazard when a drill bit breaks. The drills used are tiny and made of carbide. Carbide drills require a high-speed drill, so always wear eye protection while drilling.

Here are a few basic safety tips to keep in mind:

- Work in an area with adequate ventilation.
- Wear eye protection.
- Remove contact lenses—wear glasses if you must.
- Wear an apron or old clothes.
- Wear latex or rubber gloves.
- If working around fumes or dust, wear a filter mask.
- Be near running water in case you need to wash or flush your eyes.
- Do not mix chemicals unless directed by the manufacturer.
- Always add acid to water (don't add water to acid).
- Keep paper towels nearby to soak up spills.
- If you spill a chemical, resist the urge to react immediately; pause and think about what you will do.
- Have MSDS data for each chemical you are using before you start; you don't want to have to hunt them down in an emergency.

Terminology

One thing that makes PCB design intimidating is the specialized lingo surrounding it. Here are some terms you should be familiar with:

Ammonium persulphate   A clear liquid that can remove copper from boards.

Annulus   The shape formed by the space between two concentric circles (such as the shape of a pad around a hole).

Artwork   The design for a PCB.

Autorouter   A computer program that can make connections automatically on your board.

Computer-Aided Design (CAD)   The activity of drawing schematics and artwork using special software is known as Computer-Aided Design.

Computer-Aided Manufacturing (CAM)   Professional board manufacturers use special machines to produce PC boards. These machines accept information about the board you wish to produce using specially formatted files. The process of producing boards (or anything, for that matter) from these files is known as CAM and the files are often called CAM files.

Component side   The side of a through-hole PCB where the components are mounted.   = silk-screen side

Design rules   Guidelines for the PCB layout (for example, no two traces can be closer than a certain distance). Many programs can automatically check for design rule compliance.

DRC   Design Rule Check.

Dry film   A film used to produce the solder mask layer.

Electronic Design Automatic (EDA)   A generic term for using computers to design circuits.

Etch   The process of using a chemical to remove copper from a board.

Excellon drill file   A file that describes the size and position of holes in your board.

Ferric chloride   A dark liquid that removes copper from boards. Ferric chloride is inexpensive and safe, but it works best hot and is notorious for staining skin, clothes, and concrete.

FR4   Fire-retardant laminate made of fiberglass and resin, and used as a base for many PCBs.

✳ **Gerber file**   A file that describes the layers of the PCB.

Laminate   The insulating board covered with copper on one or two sides.

Land   A copper area meant for attachment of a component.

Liquid Photo Imageable (LPI)   A type of chemical used to create a solder mask.

✳ **Mil**   A unit of measure often used in PCB layout. A mil is .001 of an inch.

✳ **Net list**   A machine-readable file that describes a schematic.

Pad   The copper area where components are attached.

Panel   A standard-sized board that a manufacturer builds. The panel may contain multiple boards.

Panelization   The process of converting placing multiple boards (often identical) on a single larger panel. This often reduces the cost of producing the boards. The boards may be cut apart after processing or after components are mounted on them.

Photoplotter   A special machine that draws images from a Gerber file onto film using a light beam.

{ ammonium persulfate
{ ferric chloride

Resist   A chemical that prevents another chemical from attacking copper on the board.

✳ **Schematic capture**   Entering a schematic into a computer program.

Silk screen   The markings on the PCB that identify components. The process, however, is not always done with a silk screen printer.

Solder Mask Over Bare Copper (SMOBC)   This is one way of placing solder mask on a board.

Solder mask   A layer that prevents solder from adhering to a board in certain areas. The layer usually covers the entire board except for areas that should be soldered (for example, component pins). This simplifies hand soldering and may be essential for certain automated soldering processes.

Solder side   The side of a through-hole PCB where the solder is applied.

Step and repeat   A process where multiple identical boards are placed on a larger PCB (a panel).

Substrate   The insulating part of the board. It is usually fiberglass, but it is occasionally phenolic paper or ceramic.

Surface mount   Parts—usually very small parts—that solder directly to the surface of the PCB (as opposed to through-hole parts).

Through hole   Parts that have leads intended to pass through holes in the board so that the part itself is on one side of board but soldered to the other side of the board.

Tooling   Creating reusable tools used to manufacture a board.

Trace   A copper conductor on a PCB.

Track   A copper conductor on a PCB.

Via   A hole that exists only to connect one layer of a board to another layer.

Wet Film   A chemical used to form the solder mask layer.

In Summary

Modern computer software and manufacturing methods put PCBs in reach of every designer, no matter how small. With the many advantages of PCBs, you have no excuse not to make use of them, even for prototypes and small jobs.

This chapter has covered a lot of ground, but don't worry—you'll see all of this, a piece at a time, in future chapters. Just try to keep the big picture presented in this chapter in mind as you progress through the book.

Eagle and Schematic Capture

When I was in high school, I nearly flunked drafting. We were required to draw on large sheets of green paper that seemed to have a chemical reaction to any sort of eraser. Each erasure mark the instructor could detect was a letter grade off that drawing. I won't even describe the week we did ink drawings with old-fashioned ruling pens. I was relieved (and disheartened) to find out a few years later that in the real world no one actually drew on paper. (They used drafting film, which was easy to erase with a plastic eraser.) Also, they used real technical pens instead of the horrible things we used in school.

Too bad we didn't have today's computers back then. Just as computers make it easier to prepare and change documents, they also simplify drafting. Although many general-purpose drafting programs are available, you want a program that is specifically for electronics.

The CD-ROM at the back of this book contains the free version of a powerful electronics drafting program named Eagle (a product of CadSoft). The program contains three modules: one for schematic capture, another for laying out *printed circuit boards* (PCBs), and an autorouter. Eagle runs on Windows or Linux and is an excellent program.

As with any complex program, it will take you a while to get comfortable using Eagle. You may be tempted to just draw your schematic on the back of a napkin like you always do. Resist this urge. It may seem like a lot of trouble putting the schematic into a computer. However, armed with the schematic, Eagle (and most other modern *Computer-Aided Design* [CAD] programs) can check your work (and even do some of the work for you).

Before long, you'll find it easy to draw a schematic using Eagle. The results are very professional looking (see Figure 2-1). Eagle has a large standard library of components, and for now the schematics we'll draw will only use those components. However, in Chapter 5, "Custom Libraries," you'll see how you can define your own components in case you need something the standard library doesn't contain.

Introducing Eagle

Installing Eagle is pretty much like installing any other application. If you are running Windows, you simply run the setup program from the CD-ROM and answer the questions presented. When you first run Eagle, it will ask you for your license disk and key. This only applies if you have purchased a license. If you haven't, simply click the "Run as Freeware" button.

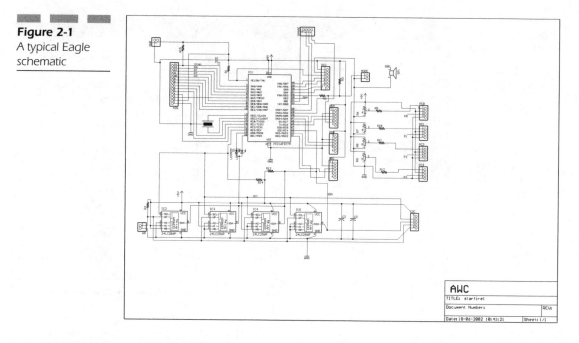

Figure 2-1
A typical Eagle
schematic

When running in freeware mode, Eagle restricts you to schematics that only contain one page. However, this is more than adequate for many projects. In particular, Eagle freeware also restricts the PCB's size as well, so you couldn't use a very large schematic anyway.

When you start Eagle, you'll see the Eagle Control panel. This screen (see Figure 2-2) lets you explore several categories:

- **Libraries** Libraries contain the components you use to build circuits (such as resistors, capacitors, and integrated circuits). In Eagle, each library can contain many components that include both a schematic symbol and an associated PCB pattern (the footprint).

- **Design rules** Eagle can check that a board meets certain design criteria known as design rules. For example, you may not want traces closer than 10 mils and Eagle's design rules can make this check. You may want different sets of design rules to differentiate between boards you make yourself and different manufacturers (who may have different capabilities). Even when using a single manufacturer, it may be less expensive to have boards produced to lesser specifications, so you may want to manage different design rule sets.

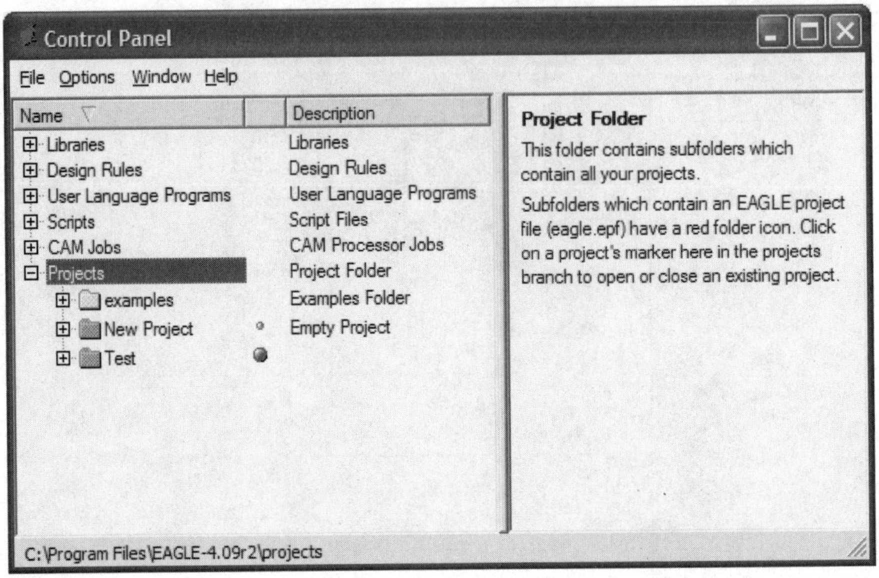

Figure 2-2
The Eagle Control
Panel

- *User language programs* (ULPs)   Eagle supports a C-like
  programming language that you can use to write custom operations.
  These are known as ULPs.
- **Scripts**   In addition to ULPs, Eagle also enables you to build scripts.
  These are just lists of commands that save you from entering the same
  sequence of commands over and over again.
- *Computer-Aided Manufacturing* **(CAM) Jobs**   Eagle can store
  CAM jobs to produce output from your finished designs. You might
  have one CAM job to produce files for an outside manufacturer and
  another for printing output on a laser printer for producing your own
  boards.
- **Projects**   This is the main part of the control panel that you'll be
  interested in for now. The Projects folder contains subfolders that hold
  your working files (one project per subfolder). Projects can contain
  schematics, board layouts, scripts, ULPs, CAM jobs, and text files.

To create a new project, select File | New | Project from the main menu.
You'll have the chance to type in a new name (to replace the default name
of New_Project_X, where X is a sequential number).

Eagle considers one project "open" and places a green dot to the right of
it in the list. When you create a new project, it automatically becomes the
current project. To switch projects, use the File | Open | Project menu or

right-click on the project you want to open and select Open Project from the resulting menu. When something (such as a schematic) is in the project, you can double-click it to open the particular file. In this case, nothing is in the project, so select File | New | Schematic (or right-click on the project and select New | Schematic).

A Schematic

Figure 2-3 shows a screen shot of Eagle's schematic editor running on a Windows PC. Most of the interface should look familiar—toolbars and menus on the edges of a main document. Also, a status area is located at the top left (just to the right of the eye icon in the vertical toolbar). This area shows the current grid setting and the current cursor position. To the right of the status area is an unusual feature—a command line.

Any command you can execute in Eagle has a text format and you can enter text directly into this line. You can also use the mouse to perform most functions, but occasionally typing precise parameters makes more sense.

Figure 2-3
The schematic editor

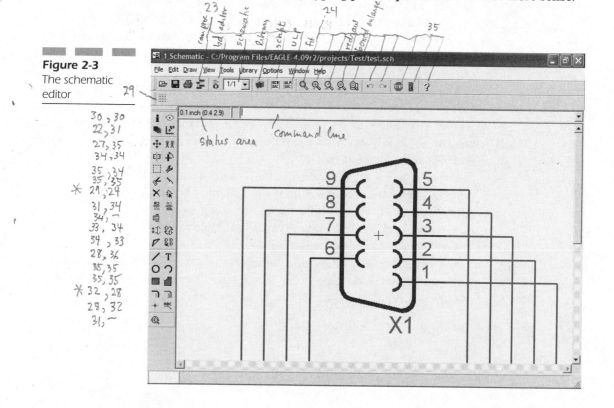

About Layers

Every drawing in Eagle is divided into layers. You can imagine each layer as a sheet of clear plastic that has certain elements drawn on it. When you stack the sheets together and look at all of them together, you see the complete schematic. Schematics have the following layers, by default:

- **Nets** Wires that connect components together.

- **Busses** To simplify schematics, you can group nets together into busses.

- **Pins** The connection points for each component appear on this layer.

- **Symbols** This layer contains the actual schematic symbols.

- **Names** This is a text layer that holds the component names (such as R1 or C1).

- **Values** This is another text layer and it contains the values of the components (such as 10K).

Each layer appears on your screen as a different color. You can use this icon ▉ to bring up the display dialog (see Figure 2-4). This dialog enables you to add and delete layers, change layer colors, and also hide or show layers. For schematics, it isn't very useful to hide most layers. You might, however, want to hide the values on the schematic to keep the screen uncluttered.

The All and None buttons on the display dialog enable you to turn all layers on or off with one button press. The All button is a useful shortcut when you want to show everything. The None button can be useful when you want to show only one or two layers. You can click None and then just click on the layers you want to see.

For now, you want all layers visible. When working with schematics, you may find layers of little use, but when designing boards, layers are crucial. Also, you want to be careful that when adding elements such as text, you place them in the correct layer (you'll see shortly that you can control the layer of objects from the toolbar).

If you accidentally place an element in the wrong layer, you can change the layer, as you'll see later. In schematic entry, many elements you add will automatically select the correct layer, so you won't have to worry.

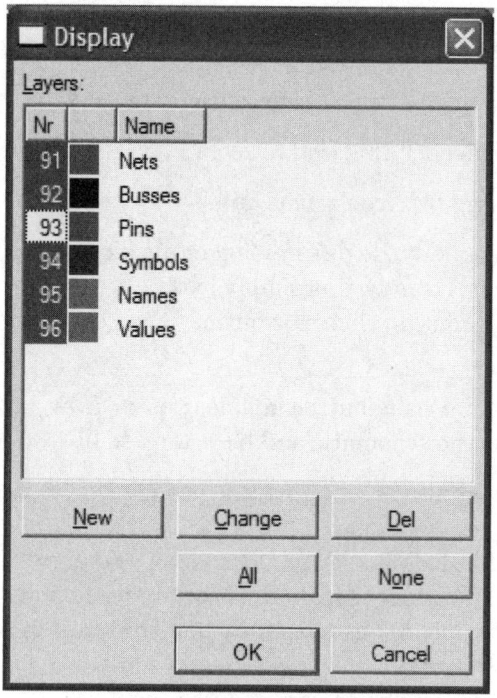

Figure 2-4
The Display dialog

Nr		Name
91		Nets
92		Busses
93		Pins
94		Symbols
95		Names
96		Values

New   Change   Del

All   None

OK   Cancel

Main Toolbar

The top toolbar stays the same no matter which tool you select. Most of these are self-explanatory. If you hover the cursor over a button, you'll see a balloon that tells you the button's function. As you might expect, the first three buttons open a file, save a file, and print the current file.

The next two buttons select the CAM processor (see Chapter 7, "Eagle Output") or the board editor (see Chapter 3, "Board Layout"). Directly to the right of the board editor button is a dropdown menu that enables you to select the schematic sheet you are viewing. Unless you have a paid license for Eagle, you are limited to one sheet, so this dropdown menu isn't useful for the free version.

The next three buttons enable you to select a library to use, run a script, or run a ULP. Of the buttons discussed so far, only the first three are useful at this point because they refer to functions you haven't encountered yet. However, the buttons to the right of the ULP icon are very useful:

This button fits the entire schematic on your screen.

You can click this icon to zoom in on the schematic.

Clicking this icon zooms out.

Occasionally, Eagle's erasing or altering an item will corrupt the display. Usually, this simply looks like gaps in lines and is harmless. You can click this button to force a redraw of the entire display.

The final zoom button enables you to draw a box around a portion of the schematic and blow it up to fill your screen.

The next two buttons (and) enable you to undo and redo operations.

The icons that look like a stop sign ( ) and a green light ( ) enable you to cancel or start an operation. These are useful with certain commands that normally apply to one element when you want to apply the command to the entire drawing.

The final icon on the toolbar ( ) permits you to ask for help from Eagle's online help system. Keep in mind that you can access all the toolbar buttons in other ways. For example, the View menu enables you to select the zoom functions. In addition, each button has a command that you can type into the command-line box just above the drawing. Also, Eagle assigns keys to most functions. You can customize these key assignments, so it is impossible to say exactly which keys map to which functions. By default, for example, F10 maps to undo and F9 maps to redo, but you can change those assignments (using the Options | Assign menu item).

Selecting Components

Naturally, the first things you need when drawing a schematic are components. You can use the Add command ( ) to open the dialog that appears in Figure 2-5. The components are organized by category. You can scan through them or use the search function to locate the part you need. If you make the dialog wider, you can also see the descriptions of each library, although they are usually hidden by default.

^{1} on LHS to right of X

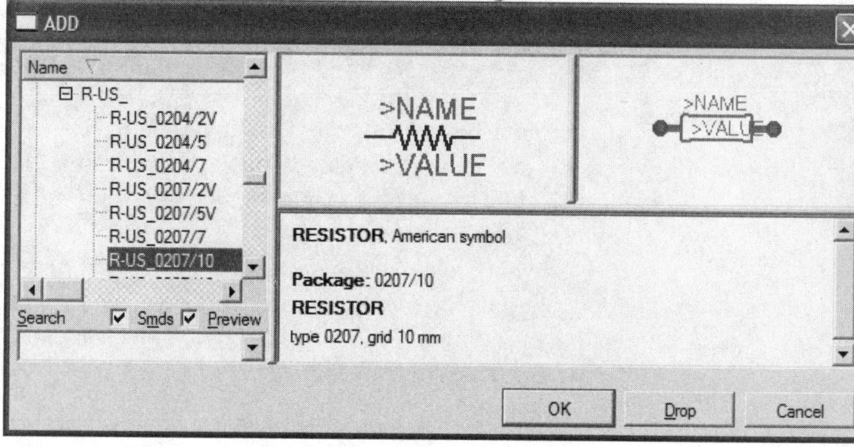

add dialogue

Figure 2-5
The Add dialog

If you haven't worked with a PCB layout program before, you may find component selection a bit surprising. When you are building a circuit on a perforated board or breadboard, you are mostly concerned with a component's electrical value—that is, a resistor's resistance in ohms or a capacitor's capacitance in microfarads. When you lay out a board (either by hand or with a computer) you are mostly concerned with the component's physical size. For example, an electrolytic capacitor may be 10 millimeters in diameter and have leads spaced 5 millimeters apart. The actual component value isn't important at the board level—just the physical dimensions.

You can usually find physical dimensions for components in the manufacturer's data sheets. Just be careful . . . parts from different vendors may not have the exact same dimensions. For example, a Radio Shack 220uF 35V electrolytic capacitor has 5-millimeter lead spacing, but the same device from Panasonic uses a narrower lead spacing.

When possible, it is best to have the exact parts you will use on hand. Another useful item is a caliper—especially a digital (electronic) caliper that is much easier to read than a regular caliper. These are quite inexpensive now (check out vendors such as Harbor Freight at www.harbor-freight.com for inexpensive digital calipers). If you don't have a caliper, you can get by with a good ruler for most purposes.

One major problem is that Eagle has so many components that it is difficult to find exactly what you need until you get accustomed to the available components. Make sure the Preview button is checked so you can see the outlines of the component. You can also uncheck the *Surface Mount Device* (SMDs) button to hide surface-mount components if you aren't using

Reading a Standard Caliper

If you don't have a digital caliper, you can still get an old-fashioned mechanical caliper. Some of these have a dial that you can read directly, but these probably cost as much as a digital one and you'll enjoy the digital caliper more.

However, vernier calipers—especially those made of plastic—are quite inexpensive. These aren't nearly as good as a professional-grade caliper, but they will often do the trick. The problem is learning to read the scales. Here are the basic steps:

1. Before measuring, make sure the caliper reads 0 when fully closed (you'll see how to read the markings shortly). If the reading is not 0, adjust the caliper's jaws until you get a 0 reading. If you can't adjust the caliper, you'll have to remember to add or subtract the correct offset from your final reading.

2. Close the jaws lightly on the item you want to measure.

3. If you are measuring something round, be sure the axis of the part is perpendicular to the caliper. That is, make sure you are measuring the full diameter.

   The caliper has a fixed scale and a sliding scale. Usually one side of the scale is marked in English units and the other in metric units. The fixed scale is marked in centimeters (or 10s of millimeters). The moving scale is marked from 0 to 9 (or sometimes 10). The nearest whole number on the fixed scale that is to the left of (or exactly on) the moving 0 indicates the number of millimeters.

3. You can use the marks on the moving scale to read down to 0.1 millimeters. The first mark (reading from the left) that lines up exactly with a mark on the fixed scale indicates the remaining digit.

   You can actually read more digits from a quality caliper, but for our purposes, 0.1 millimeters should be enough.

A common caliper has jaws you can place around an object, and on the other side, jaws made to fit inside an object. These secondary jaws are for measuring the inside diameter of an object. Also, a stiff bar extends from the caliper as you open it that can be used to measure depth.

them. You can also browse the available libraries from the control panel if you prefer.

Components have two parts: a schematic symbol and a package. The symbol is what you see on the schematic. The package is what appears on the final printed circuit board. That means that the same component may have many entries (one for each possible package). For example, no one symbol for a resistor exists. Instead, a symbol for various sizes of resistors in both horizontal and vertical orientations is used. Symbols are also used for surface-mount and through-hole resistors. To further complicate matters, certain symbols use the American-style schematic symbol and another set uses the European-style symbol.

If you open the RCL folder in the Add dialog, you'll notice it has several subfolders. The R-EU_ folder contains European-style resistors, and the R-US_ folder has the American symbols. Opening the R-US_ folder reveals dozens of separate resistors. Which one do you need? For the schematic, these are all interchangeable. Your choice depends on what you want when laying out the PC board. For example, R-US_0204/2V is a type 0204 resistor (1/8-watt resistors are usually type 0204). The resistor has leads spaced on a 2.5-millimeter grid and in a vertical orientation. Compare this to the R-US_0204/5, which is a resistor of the same size on a 5-millimeter grid in a horizontal orientation. A standard 1/4-watt resistor is a type 0207. The folder also contains surface-mount resistors such as R-US_M3516. On the schematic, these parts all look identical, but they are quite different on the finished board.

Once you select a part, the cursor appears as the schematic symbol. You can left-click the schematic to place the component at the desired location. You can repeat this as many times as you like, so if you need five resistors, you can put them all down with just five clicks. Right-clicking rotates the component 90 degrees. You can keep right-clicking to rotate the component's cursor until the component is in the orientation you want. The new orientation will apply to subsequent left-clicks but won't change any components you've already placed. You can stop placing the component by pressing Escape (or the Stop sign on the toolbar).

Here is a common idea in Eagle: First, you select a command and then you click on a location. Compared to many Windows programs, this style of operation may seem backwards. For example, in Microsoft Word you select a position and then a command. Eagle is just different and you have to get used to it.

If you don't like the position of the components you've already placed, you can use the Move button (⊕) to alter any component's location. You simply click the toolbar button and then click the component you want to move.

Each component has a small crosshair (usually in the middle) you click on to select it. You don't need to hold the mouse button down. Just let go and move the mouse to the new location. Another click will release the component. Right-clicking will rotate the component just like it does when you originally place the component. You can continue moving components until you select another tool. Moving a component that has connections will cause the connections to remain connected and stretch to fit the new position.

If you have components that are close together, Eagle may not be able to tell which component you want to move. Anytime Eagle can't decide where you meant to click, it will prompt you. Eagle randomly picks one component near your click and turns it red. Then in the status bar near the bottom of the page, you'll see this message:

Select highlighted object? (left = yes, right = next, ESC = cancel)

If the red item is not what you meant to select, simply right-click until the correct item turns red. Then left-click to select the item you want.

Making Connections

Components aren't much good without connections between them. A common mistake is to use the Wire command ( ) to connect components. Don't do this under any circumstances! It seems to make sense to use wires, but in reality, you want the Net command ( ) instead. Click the Net command icon and then click a component's terminal. You can drag the mouse around to move the line and left-click to make a bend in the net. At the top of the screen (below the main toolbar), you'll see buttons that enable you to define which angle of bend you want to make. You can also cycle between the bend angles by right-clicking. When you click another component terminal, the line ends. You can also double-click to end the net early.

Draw nets

Keep in mind that the nets represent conductors on the final printed circuit board, but their lengths and bends have no effect on the PCB layout. The lines on the schematic are to indicate, simply for your benefit, the connections required by the circuit. The physical layout is not a concern on the schematic.

@27

You can move parts of nets with the Move command ( ). The net will stay connected. When you connect a net to another net (except at the edges), the Net tool automatically inserts a junction. You can also manually insert junctions using the Junction tool ( ).

Net Classes

Another item on the secondary toolbar enables you to select a net class. Usually, all your wires will be of the default class. In the Edit | Net Classes menu item, you can bring up a dialog that enables you to define new net classes. Each class can have a custom width, clearance, and drill size.

The most common reason you'll use net classes is to define power and ground connections. You may want to make these traces wider than other traces to improve the current-handling capacity and noise immunity. These widths won't be used until you create a PCB, but you'll want to define them at the schematic level so that when you create the board, the correct parameters are applied.

The Grid

Eagle drawings operate on a grid. Every click of your mouse is lined up to the nearest grid point (which is very handy). You can modify the grid properties using the View | Grid menu item or the Grid Toolbar button ( ). You can turn the grid visibility on and off, change the grid to appear as dots or lines, and set the distance between each grid point.

I personally prefer to have the grid invisible, but you may prefer to have it on. One caveat: It is important to have libraries that line up on the grid size you are using. If you have a library drawn on, say, a .25-inch grid and you set your grid to .1 inch, you may have trouble attaching nets to components.

Practice

To practice what you've learned so far, try reproducing the schematic in Figure 2-6. This is a DB9 connector (the F09H connector from the con-subd library) and a 9-pin header (PINHD-1X9 from the pinhead library).

Remember, you can use F2 to redraw the screen. Also, if you make a mistake, you can use Undo (Alt + Backspace or F10) or you can click the delete icon (✗) and then click on an element to delete. It doesn't matter if you have the nets exactly the same as the figure, as long as they all make the correct connection and the nets don't run together so that you can't separate them visually.

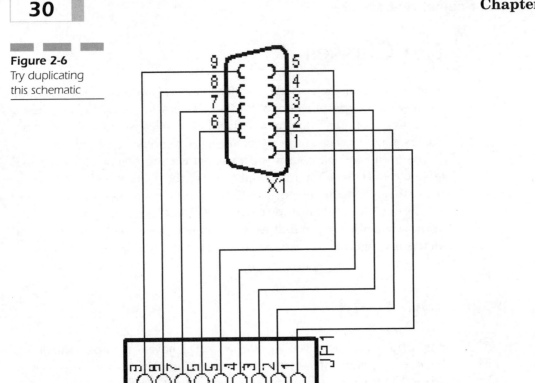

Checking Your Work

You can check your work and get information about schematics that you have open in several ways. The Show command (⊙) highlights a connection on a component (a pin) or a net and all the nets and pins that connect to it. If you click the Show command and then click a net, it should turn bright green, and the pins it connects to should turn bright red. If you have any disconnects, they won't become brightly colored. In addition, you shouldn't see any extra parts of the circuit light up. Also, this command shows the name of the net in the status bar.

If you have nets that are close together, be sure Eagle isn't just highlighting the net to ask you if it is the one you mean to select. That will make only a portion of the net bright green, which can mislead you into thinking the net is not connected as you expect.

Another way to get information about a schematic is to use the Info command (**i**). Clicking on this icon and then clicking a schematic element will show you information about the item (see Figure 2-7). Notice that to

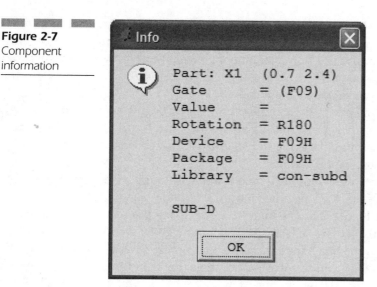

click on a component, you have to click on the crosshairs (usually near the center of the part).

The best check you can perform is an *Electrical Rule Check* (ERC) (⊕). This process shows any errors or warnings that Eagle detects. If no errors occur, you'll see a message to that effect in the status bar.

You may want to measure certain distances while drawing (although this is usually more useful when laying out the board). The top-left part of the screen shows the current grid size and the distance between the default reference point and the cursor (the first number is the horizontal, or x, distance; the second number is the vertical, or y, distance). You can set a new reference point ( ) to anywhere on the drawing. When you do, a second distance display appears next to the original one that shows the relative distance from your custom reference point and the cursor (prefixed with an R).

Bussing Connections

With a complicated schematic, it is often confusing to draw nets to every possible connection. Eagle doesn't really care if you connect all the parts of a net or not. All nets that have the same name are connected even if they don't appear to touch. You name a net with the Name command ( ).

Allowing disjointed nets to connect can be useful when you have a net that goes everywhere (and is grounded, for example). To make it more obvious, you can use the Label command (🔤) to place text near the net that reflects its name. While placing a label, you can right-click to rotate it the same way you rotate components.

If you don't assign nets names, they receive default names (such as N$1, N$2, and so on). For an ordinary schematic, it isn't a problem to retain these names, but if you are going to use implied connections, you'll want to give the nets more explicit names.

Another way to simplify connections is by using a bus. You draw a bus with the ⌐ tool. Using a bus requires a bit of name planning.

Look at Figure 2-8. The bus is named (using the Name ⧉ command) Bus1:Pin[6..9]. This tells Eagle you intend to connect Pin6, Pin7, Pin8, and Pin9 to the bus. When you connect a new net to the bus, a menu pops up that enables you to select one of the signals on the bus. Keep in mind that Pin is just a net name. It might just as well be Bus1:X[6..9] or Bus1:DC[6..9].

Figure 2-8
A schematic with
a bus

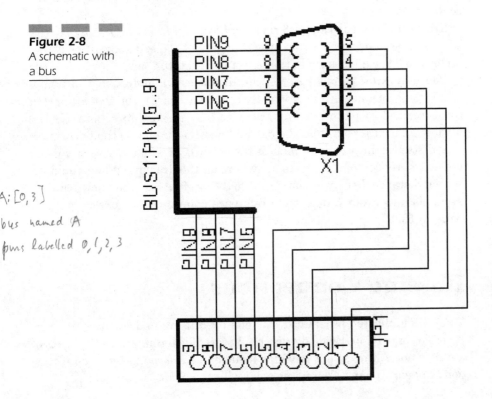

A: [0,3]

bus named A

pins labelled 0, 1, 2, 3

If you have existing nets, you'll have to name them to match a signal on the bus before you connect them to the bus. Bus connections don't affect the PCB layout, but they do make your schematic easier to read.

Power and Ground

Power and ground connections are often treated specially in Eagle. Many components have predefined power and ground pins that will automatically connect to the appropriate net. Normally, Eagle won't show these connections.

For example, consider a 7400 integrated circuit. In addition to the input and output pins, Eagle knows the chip requires a power (Vcc) and ground (GND) pin and will connect them to the chip. The only problem with this arrangement is that you may have a mix of parts that require different power nets. For example, you may have some components with Vcc connections and some with Vdd connections.

invoke

One potential solution is to simply draw a Vcc net and a Vdd net and then connect them together. However, a better answer is to use the invoke command ( ) and select the component. This will show you any gates and pins not currently on the schematic. Then you can add any of these to the schematic. Once the power and ground nets appear on the schematic, you can connect them as you see fit instead of depending on Eagle's automatic decisions.

To correctly model the supply nets, you have to connect them to a component from the supply1 or supply2 libraries. These don't resolve into PCB components, but they do let Eagle know which power supply net is which.

Invoke also enables you to select individual parts of components that have multiple subcomponents. For example, a 7400 integrated circuit has four NAND gates in one package. When you add a 7400 to your schematic, only one gate appears. To add any of the other three gates, use invoke.

Another trick you can play with gates similar to the 7400 is to swap pins and gates. This is mostly useful when laying out the board, but you can do it while drawing the schematic. Suppose you have a 7400 gate where input net A connects to pin 1 and input net B connects to pin 2. When it is time to lay out the board, it might work better if A connects to 2 and B connects to 1. Functionally, no difference occurs, and that's what pinswap ( ) does. A simple resistor or capacitor is another example of a component where the pins are interchangeable. When a device (like the 7400) has multiple identical gates, it doesn't really matter which gate is used for which set of connections. So, if it makes for an easier board layout, you could just as well connect A to pin 4 and B to pin 5. Then the output would be on pin 6 instead of pin 3, but otherwise no difference exists. Eagle can make this swap by executing the gateswap ( ) command. To swap a gate or a pin, select the correct command and then click the first item you want to swap. When you click the second item, the two items will switch places.

Editing

One of the advantages of drawing schematics on a computer is that you can make changes. Several tools enable you to change your drawing. The Mirror command ( ) lets you create a mirror image of a component. This command flips a component in a different way than rotating it. For example, suppose you place a 1×6 pinheader. By default, the connection points (that look like tails on each pin) will point to the left, and pin 1 will point to the top of the schematic. If you rotate the part 180 degrees, the tails will point right and pin 1 will be toward the bottom of the page. However, if you mirror the default component, pin 1 will remain at the top, but the tails will flip to point to the right.

You can rotate a component using the rotate command ( ). Of course, you can also move the component and right-click to rotate, which is usually just as easy.

Sometimes you will want to put an extra corner in a net. You can do this with the Split command ( ). This inserts a bend in the net and enables you to set the wire just as if you were inserting it for the first time. That means you can drag it around and right-click to change the bend shape.

The ultimate editing tool, however, is the Change tool ( ). This versatile tool enables you to change an item's layer, width, style, size, font, ratio, text, roundness, class, package, or technology. Not all these modifications apply to every sort of element, of course. (In addition, some of these modifications apply to items you haven't read about yet.)

As an example, suppose you draw a net and then decide you don't want to use the default class for it. You can simply click the Change tool, select class, and then pick a new class from the resulting list. Then you simply click on the net (or nets) to change.

You can change the name and values of components using the and tools. Sometimes the default position of these labels isn't where you want them. In that case, use the Smash ( ) command to separate the labels from the components. Then you can move the labels and manipulate them as you would any other component.

Working with the Clipboard and Groups

Eagle lets you define a group using the Group ( ) tool. Simply draw a box around the elements you want to include in the group. You can drag a rectangular box or click points to define an arbitrarily shaped polygon. With a group defined, you can use the Cut command ( ) to put the selected items in the paste buffer (which is equivalent to the clipboard). Once you select the Cut tool, you have to click somewhere on the schematic to initiate the cut. You'll get an error if no group is present. The point you click will become the reference point Eagle uses when placing the group in a future paste. If you want the reference point at the center, just click the Go ( ) icon on the toolbar.

Unlike most Windows programs, Cut doesn't remove the group from your drawing, so it is similar to the Copy command in most programs. You can paste the buffer into your schematic (or a different schematic) using the Paste command ( ). Then you can place the group as you would place any component.

Many other commands will work with groups. The key is to use the right mouse button to apply the command to the group instead of an individual element. For example, if you want to delete a group, select the delete icon and then right-click anywhere in the schematic. Most commands that operate on an individual component can also work on a group by simply substituting the right mouse button for the left one.

If you just want to copy a single component, you can use the Copy ( ) command. Simply click the tool and then click the component you want to copy. The cursor will change just as if you had used the Add command.

Drawing Additional Items

You can add arbitrary shapes to your schematic using circles, arcs, wires (lines), rectangles, and polygons. Like other items, you simply select the appropriate toolbar button and follow the directions in the status bar.

These can be useful for annotating a schematic. For example, you might draw a dotted line to divide your schematic into two parts. You can also use the shapes to draw logos, for example.

It is also possible to draw arbitrary text on any layer of the schematic. When you insert text, the secondary tool bar will show a variety of options. You can do the following:

- Set the layer for the text.
- Rotate or mirror the text.
- Set the size of the text.
- Set a ratio for the text (see the following text).
- Change the font.

The size of the font is measured based on the current grid settings. If your grid is set to inches, the font size is in inches. The width of the font is set by the ratio parameter. If you set a font size to .3 inches and the ratio is 10 percent, the width of the font's lines will be .03 inches.

Frames

One way to spruce up your schematics is to add a frame from the frames library. If you are printing on regular-size 8.5-inch by 11-inch paper, the LETTER_L and LETTER_P frames will place a landscape (horizontal) or portrait (vertical) frame on your drawing. The frame is just another component, but it has a border and a name plate that you can fill in. This makes your drawing look more professional but doesn't affect the finished PCB at all.

You can use the text tool to customize the frame. Later, you'll learn how you can modify a frame to have your own text in it and then save it in a custom library for reuse. We will discuss that in Chapter 5.

Output

Once you've finished your schematic, you will probably want to print it, which can be done via the Print dialog (accessed by going to File | Print). Here you can scale the drawing's size and select a variety of options. For example, you can select several rotation and mirror orientations.

The other way to generate output from your schematic is via File | Export. This command enables you to export in several formats:

- **Net list**   A text file that describes all the nets in your schematic. The file contains each net, its class, and the components and pins that connect to the net.

- **Net script**   This is an output format similar to a net list, but it is in the form of an Eagle script file. Eagle can run this script to insert the nets in a board design.

- **Part list**   When you export a parts list, you create a bill of materials that contains the parts in your schematic and information like their value and package.

- **Pin list**   A pin list shows each component and the connects made to each pin.

- **Image**   You can export the schematic as an image in a variety of file formats (including bitmap, jpeg, and png). You can save to a file or put the data on the clipboard. You can also select the image resolution and if you want color or monochrome data.

In Summary

At first glance, it doesn't seem like schematic drawing has much to do with making PCBs. However, as you progress to the next chapter, you'll discover two important things. First, having a schematic lets Eagle check your board work and also lets it do some of the work. Second, most of the commands that you use to draw schematics are exactly like the ones you use to draw PCBs. Sure, a few differences exist, but for the most part, you work with boards in the same way you work with schematics.

It is possible to use Eagle without drawing a schematic. However, schematics are much easier to check for errors and change. Generating a board from a schematic is easier and will reduce errors.

Armed with your schematic, you are ready to lay out a board. In the next chapter, you'll start working with the board editor to produce a layout.

Board Layout

Snake was the name of an old-fashioned video game. The idea was you guided a snake (well, just a big white squiggle on a black and white monitor) around to eat some sort of targets. That sounds simple, but it had a catch. The snake kept growing longer and longer. If the snake hit his own tail, the round was over.

Sometimes it seems like laying out a *printed circuit board* (PCB) is similar. Your goal is to guide these copper traces around the board to connect the various nets defined on the schematic diagram, but the game has a catch: The traces are not allowed to cross on the same layer.

About Back Annotation

One of the major features of Eagle is *forward and back annotation*. This simply means that the changes you make on the schematic are reflected on the matching board and vice versa. A few caveats exist. First, you must keep both documents open at the same time. When you open a schematic or board file, Eagle will ask for permission to open the other file automatically. If you don't open the files together, Eagle will warn you that it won't perform back annotation. If the board and the schematic get out of synch, you'll be sorry, so just keep both files open.

The other problem is that it is possible for an error or problem in one file to break the link between the files. This is one reason that backup files are very important when working with Eagle. Eagle automatically saves a backup each time you save your work. In the Control Panel, the Options | Backup setting will let you set the maximum number of backups to keep and can even automatically save backups every few minutes.

The backups are named with a file name similar to the original. For schematics, the file extension of the newest backup will be .s#1. The next backup will be .s#2 and so on. Timed backups will have the .s## extension. The corresponding board files will have extensions of .b#1, .b#2, and .b##.

If you ever need to recover a backup, you should copy the original file (just in case). Then rename or copy the backup file to have the name of the original file. So if you want to recover the latest backup of your board file (and it was called mydesign), you'd copy mydesign.sch to mydesign.bad and then copy mydesign.s#1 to mydesign.sch. Of course, you shouldn't have any of the files open in Eagle when you make the copies.

It is possible to use the board editor without a schematic. Some of the commands in the board editor are geared toward pure board layout. If you try to use these commands while annotation is active, Eagle will ask you to return to the schematic and make the change there.

Getting Started

If you already have a schematic, it is easy to start a new board file. Simply select File | Switch to Board. This normally just jumps to the board editor, but if the board doesn't exist, Eagle will offer to create it for you automatically.

Eagle will create a default board outline for you and also place components for each required part on the screen. In addition, each net will be represented by an *air wire*—a wire that directly connects the components. These wires will take the most direct route between points and won't be suitable for the actual layout. Your job will be to replace the air wires with real traces.

Figure 3-1 shows the default starting point when creating a board from the simple schematic in the last chapter. Of course, the parts are just clumped together outside the board area—you'll have to move them.

The board editor looks quite a bit like the schematic editor. One obvious difference is that the board has many more layers by default. The commonly used board layers include the following:

- **Top**   This layer holds the copper traces on the top of the board.
- **Bottom**   This layer holds the copper traces on the bottom.
- **Pads**   This layer contains the copper portions where components are mounted.
- **Vias**   Vias are holes in the board for the sole purpose of connecting layers of the board together.
- **Unrouted**   Eagle keeps track of nets that don't have corresponding copper traces in this layer.
- **Dimension**   The dimension layer contains an outline of the board's perimeter.
- **tPlace, bPlace**   Each component has an outline or silhouette that appears in this layer and is usually used to generate the silk screen. The tPlace layer is the top side outline and the bPlace is for the bottom layer.
- **tOrigins, bOrigins**   Each component has a crosshair used to move it. These markings appear on this layer for the top (tOrigins) and bottom (bOrigins).
- **tNames, bNames**   Names assigned to components (such as R1 or IC3) appear on this layer for the top (tNames) and bottom (bNames).
- **tValues, bValues**   Values assigned to components appear on this layer for the top (tValues) and bottom (bValues).

Figure 3-1
An automatically
created board

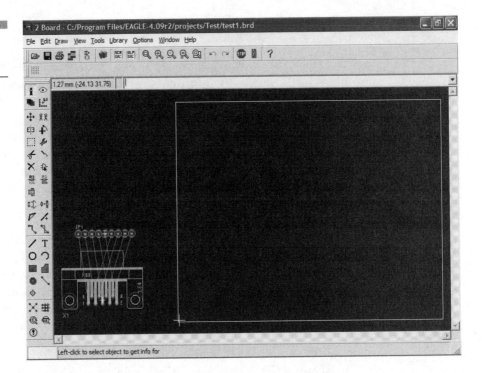

- **tStop, bStop**   These layers hold the solder mask information for the top and bottom of the board.
- **tCream, bCream**   Solder cream information for the top and bottom of the board.
- **tKeepout, bKeepout**   These layers mark areas where nothing should be placed in the area. Drawing a keepout area over a component's outline allows the design rule check to detect components that will overlap.
- **tRestrict, bRestrict, vRestrict**   These layers hold areas that the router should not use for the top, bottom, and vias layers.
- **Drills**   This layer defines holes drilled into the board.
- **Holes**   This layer defines any holes in the board.
- **Milling**   Areas (like slots) requiring milling appear on this layer.

Typically, the top, pads, and vias layers make up the top copper of the board. The bottom copper layer consists of the bottom, pads, and vias layers. Most often, only the top of the board has silk screening that is from the dimension, tPlace, and tNames layers.

Board Outline

You'll want to adjust the board outline. The outline is comprised of wires, but they are wires in the dimension layer and don't represent actual conductors. You can move them as you'd move any other lines. Although it is possible to produce nonrectangular lines by drawing arbitrary polygons in the dimension layer, you should avoid it. Having oddly shaped boards produced will drive your production costs up.

It is convenient to place the bottom-left corner of the board near coordinate 0, 0. True, you can always set an arbitrary reference point, but because the free and standard versions of Eagle are size limited, you'll get the most area to work with if you start the board at the zero point.

In many cases, it is no cheaper to make a single-sided board because the board maker probably doesn't stock single-sided blanks and will just remove all the copper from a two-sided blank during processing of your single-sided board However, if you can decide the size of the board, you'll probably expand it to be larger than you suspect you'll need and then work in one corner. Toward the end of the design process you can shrink the board outline to make the smallest possible board (or expand it if your guess is incorrect). Remember, the smaller the board, the more economical it is to produce, so you definitely want to shrink the board as much as possible.

Layer Selection

One important decision is how many layers you want your board to contain. If you are producing your own boards, you'll want to try to stick to a single-layer design. Although it is possible to produce two-layer boards, a single layer is much less trouble.

However, if you are having boards made professionally, you should go ahead and opt for two-sided boards. In many cases, it is no cheaper to make a single-sided board because the board maker will just remove all the copper from one side of a two-sided board anyway. In addition, you can usually make a smaller board on two sides, and the small board area will save more than going to a larger single-sided board.

A two-sided board has other advantages. Because each hole has copper on both sides (and plating inside the hole), the hole is stronger than a hole on a single-sided board. It is also harder to heat such a hole to the point where it delaminates from the board. Of course, you can still damage a board with too much soldering heat, but a double-sided board is harder to damage.

The biggest advantage to a two-layer board, naturally, is that you can run traces on both sides of the board. This allows you to avoid situations where you need to cross traces. Of course, on a single-layer board you can use physical wires as jumpers anywhere you simply can't route without crossing traces. For some high-speed or radio-frequency designs, you may want to make an entire layer that is mostly a ground plane to improve noise rejection.

With some versions of Eagle (but not the free version), you can create boards with multiple layers. This is a specialized technique and beyond the scope of this book. However, if you need this type of board and you understand how to create a two-layer board, you won't have any problems creating boards with multiple layers with the commercial version of Eagle. You will certainly need to have boards with three or more layers built by a commercial manufacturer.

Board Editor Commands

Most of the commands on the toolbars in the board editor are identical to those in the schematic editor.

Replace

Sometimes you want to change a component so that it uses a different package. The replace command allows you to, for example, change a TO-92 transistor for a TO-220 transistor. You select the new part and tell Eagle if you want to keep connections based on pad names or physical locations.

Optimize

The optimize command joins segmented wires that are connected together.

Route

You can use the route command—discussed in detail later in this chapter—to assign actual copper traces to nets on your circuit. While routing, Eagle keeps track of the net you are working with. You can control the width of the track and the layer.

Ripup

Sometimes you route a net and then change your mind. You can use the Ripup command to undo a route you've already routed.

Signal

If you aren't using a schematic, you can use the signal command to define a net directly on the board. If you are using a schematic (and you should), Eagle will warn you that you can't perform this operation on the board. You should switch to the schematic and define nets there.

Via

You can manually add a via (a hole strictly for connecting layers) with this command. However, when routing, switching layers will automatically place a via for you. Also, the autorouter (see Chapter 4, "Autorouting") will place vias for you automatically.

Hole

Sometimes you'd like to add holes in a board (for example, for mounting holes). This command will enable you to do that.

Ratsnest

As you route your board, you may route some nets in a way that doesn't match Eagle's idea. This is harmless, but it can make it hard to see what's left to do. The Ratsnest command cleans up the unrouted wires. The Ratsnest command also affects polygons, as you'll see later. One other handy feature: A Ratsnest reports the number of airwires remaining. When it reports 0 airwires (in the status bar at the bottom of the screen), you are done routing.

Auto

The Auto command starts the autorouter, which you'll read about in Chapter 4.

Design Rule Check (DRC)

The *Design Rule Check* (DRC) command checks your board against a set of rules that you define. The rules check to make sure that traces aren't too close to each other or to the edge of the board, for example.

Errors

The Errors command highlights any errors discovered by the DRC command.

Place Components

Once you've set a rough outline for the board, you'll need to use the Move command to drag all the components to locations on the board. Notice the airwires will move along with the part. This can be very useful. If the airwires are twisted looking (see Figure 3-2), try moving the part or rotating it (right-click the mouse) to see if the airwires get less tangled. Sometimes this is a judgment call. In general, the better the airwires look, the easier it will be to route the board. Eagle normally rotates components in 90-degree increments. It is good practice to keep components parallel to the board's edges anyway, so you shouldn't need to rotate the components to other angles.

Another good practice is to put polarized components (like *light-emitting diodes* [LEDs], electrolytic capacitors, and integrated circuits) that are along the same axis so they are oriented in the same direction. That is, all the vertical integrated circuits should have pin 1 pointing the same way. The horizontal integrated circuits should also point one way. For example, pin 1 might always be toward the top or the left of the board. This makes assembly easier and reduces errors.

Some components must go along an edge (a DB9 connector, for example). However, you don't want to put anything too close to the edge. Remember,

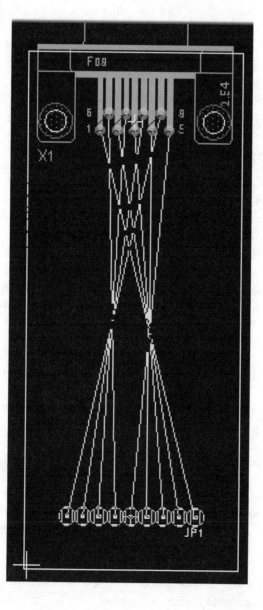

Figure 3-2
Twisted airwires
indicate an
awkward parts
layout.

a board maker will actually make several of your boards on one large panel. The more precisely the board has to be cut, the more expensive it will be to make boards.

One thing you have to watch for is clearance. For example, you might place a 7805 in a TO220 case on the board right next to an electrolytic capacitor. Eagle will dutifully comply, but when the boards are made, you find out that adding a large heat sink to the 7805 doesn't give you any room for the capacitor. (Or the capacitor takes up the space meant for the heat sink—it just depends on which one you install first.)

One trick is to get the components laid out the way you think will be acceptable. Then print the board layout (you can turn off the unrouted layer to cut down on clutter). Next, you can paste the print out to some stiff cardboard and make a mock board. With a hobby knife, a large needle (to poke holes), and scissors, you can make a mock-up board and stick sample components in it. This will tell you if you have any nasty surprises while you still have time to fix them. You'll also find any component silhouettes that don't match your actual components. It is tempting to skip this step, but don't do it. Nothing is more annoying than getting 100 boards that don't fit all of their components.

Another thing you should do with the mock-up is examine the silk-screen layer. Sometimes you'll put a marking in an awkward place and this will appear on the mockup. For example, you might mark ground on a power connector, but when the connector is installed it covers up the marking. You don't necessarily need to fix this right away, but make notes so you can go back at the end and clean up any odd silk screen.

Of course, you may have to move some components again later. If so, consider doing another mock-up. It is well worth the effort and saves future frustration.

Another thing to think about while laying out components is mounting holes. These are easy to forget because you are probably focused on the circuitry. You can use the hole command to place some mounting holes wherever you seem to have room, or you may need to place them in specific areas of the board to fit some existing holes somewhere.

Keep in mind that the size of the hole (and therefore the size of the machine screw that fits it) should relate to the weight of the board and the forces that it will encounter. A #4 machine screw is thinner than a #8 and will not be as strong.

A #0 screw is 0.06 inches in diameter (that's thin). Each size thereafter increases by 0.013 inches. So a #4 screw is 0.1120 inches. No standard screws come in sizes #7, #9, or #11. After #10, the usual way to specify screw sizes is by the diameter (such as 1/4 of an inch).

One other consideration is if you are having your boards made in another country. Although most Americans use the English measurements, a board house in another country may well use metric drills that are slightly different from the sizes you specify. Check with the board house and plan accordingly.

Surface-Mount Consideration

If you are working with surface-mount components, you have another consideration. Surface-mount components can go on the top of the board or the bottom. To flip sides, use the Mirror command.

If you are not accustomed to surface-mount technology, you may shy away from it. However, you can actually work with many surface-mount devices without much trouble at all. You do need fine solder, a small soldering iron tip, tweezers, a strong light, and magnification.

If you make your own boards, using surface-mount components has one major advantage: no holes to drill. Surface-mount parts mount directly on pads. That simplifies board construction. It also makes surface-mount boards cheaper to produce because fewer holes in the board require drilling.

Of course, you probably can't hand-solder extremely fine pitch devices and special packages such as a ball grid array. Besides, complex packages also require extremely flat boards that can be hard to produce yourself. However, if you are using a commercial board producer (and especially if you will also have the boards professionally assembled), you'll find that surface mount saves money. In addition, many modern parts are no longer available in through hole packaging.

Keep-Out Areas

Before you start routing, you may need to mark areas of the board where you don't want conductors to appear. For example, you may have to allow for a metal bracket or heat sink that would short-circuit any traces under it.

You can use the Rect, Polygon, and Circle commands to draw in the tRestrict, bRestrict, and vRestrict layers. The tRestrict prevents traces from appearing in the top layer. The bottom layer corresponds to the bRestrict layer. The vRestrict layer prevents through-plated holes (including vias) from appearing in the areas you specify.

Routing

With the components in place, mounting holes on the board, and any restricted areas defined, you are ready to start placing traces on the board.

In the next chapter, you'll see how Eagle can do this task for you—sometimes, anyway. However, sometimes you'll have to route manually (or, at least, route some part of the board manually).

To begin routing, click the Route tool. The secondary toolbar will enable you to select several options (from left to right):

- The layer you want to use
- The type of angle to use for corners
- The width of the track
- The shape, size, and drill size for any vias created

Figure 3-3 shows a board ready for routing. The board outline has shrunk, and the parts have been rotated to minimize twisting in the airwires. Figure 3-4 shows one way the board could be completely routed.

When you click on an airwire, it will change color to reflect the layer you are working with and allow you to drag it to connect it to another point on the board. When you complete the connection to a point, the process will stop until you select another airwire. You can double-click to end the trace early. Right-clicking changes the angle of bends.

You won't need it for this simple board, but you can also switch layers in the middle of drawing. This will automatically place a via to connect the top trace to the bottom trace. Ideally, you want to minimize the number of vias in your finished board, but most boards will need some. You can also use the Via tool to add vias manually.

Be careful when routing angles and corners. Keep angles at 90 degrees or greater. Making sharp angles can form an "acid trap"—a board area where etching chemicals can't flow freely. This can cause production problems.

When making corners, you'll commonly see square corners (such as the corners around pad A in Figure 3-5). However, filleted corners (the ones around pad B) are preferred at radio frequencies. For most circuits, the square corners are all that is necessary. However, Eagle does make it easy to fillet corners, so you may want to get in the habit of producing them.

If you are the circuit designer and you are laying out the board, you can often make changes on the fly that will simplify the layout. Using the PinSwap and GateSwap commands, you can reorganize the circuit to avoid routing bottlenecks. For example, suppose you have a resistor. You don't really care which side of the resistor connects to which part of your circuit. The resistor's leads are interchangeable. Another example is a 7400 NAND gate. For a particular gate, it doesn't matter which input is connected to which signal (that is, A NAND B is the same as B NAND A). You can use PinSwap to flip the inputs (or the resistor's connections) until you find the

Figure 3-3
A board ready for
routing

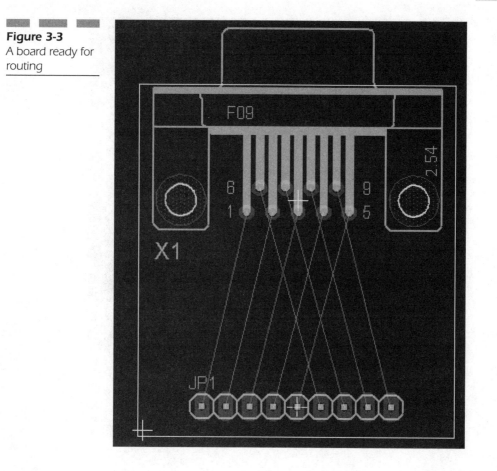

best way for routing. Likewise, four identical gates are in the integrated circuit. You don't really care which gate you use as long as each related pair of inputs goes to a single gate. The GateSwap command allows you to flip these identical gates to ease routing.

You continue routing airwires until none are left. Remember, the Ratsnest command will tell you how many airwires are remaining.

Trace and Drill Guidelines

This seems simple enough, but you need to be aware of a few issues. In particular, you want to be careful about selecting the trace width. In general,

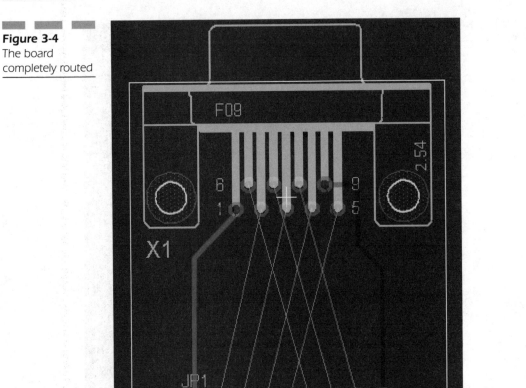

Figure 3-4
The board
completely routed

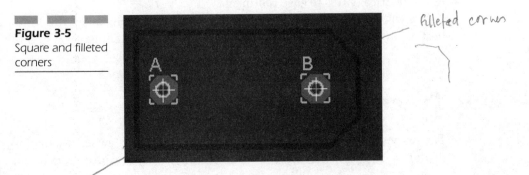

Figure 3-5
Square and filleted
corners

filleted corners

square corner

you want to make the traces as wide as possible, subject to several constraints. First, the smaller the traces, the more the board will cost. A board with 8 mil traces is reasonable for most board makers. A 10 mil board is even better. If you are making your own boards, you want even wider traces. Even if your board maker can handle small traces, you still want to make them as wide as possible.

In addition, traces must be spaced away from other elements (including other traces). This is a function of the board maker, but 8 or 10 mil separation is common. Again, if you are making your own boards, you want to have as much space as possible between elements to simplify the production.

If your board house can handle it, you probably want to stay with 8 mil traces and 8 mil separation (minimum). If you are making boards by hand, you won't want to go below 12 mil traces and spacing, and more would be even better.

Another issue is how much current and voltage a trace will carry. Traces that carry more current need to be wider. Traces with higher voltages need greater separation from other traces.

In actuality, the current-handling capacity is proportional to the cross-section area of the trace. That means you need to know the thickness of the copper on the board. Typical boards are rated as 1 ounce (35 uM) or 2 ounce (70 uM). As a guideline, you should use the values in Table 3-1 as the maximum.

With spacing, the critical factor is the maximum voltage difference between conductors. For example, a 5V supply trace and a 12V supply line have 7V between them. The exact spacing depends on several factors—the voltage differential, any coating (like solder mask) in use, and even the board's operating altitude.

Table 3-1

Current-handling capacity

Conductor width	1 oz. copper	2 oz. copper
5 mils	500mA	700mA
10 mils	800mA	1.4A
20 mils	1.4A	2.2A
30 mils	1.9A	3A
50 mils	2.5A	4A
100 mils	4A	7A

For uncoated boards, up to 10,000 feet allow 15 mils between traces or .2 mils per volt, whichever is greater. Solder mask allows closer spacing (but remember that some of your board won't have solder mask).

Usually, you want to make your power and ground traces wider than other traces—they will carry all the current. A wider ground will also provide lower impedance for noise, which is a good thing.

Setting net classes is a great way to handle power and ground traces. If you create separate classes for traces with special width and spacing requirements, it helps you remember which traces are special. Also, the autorouter will respect the class settings (see Chapter 4).

Pads should be wider than the traces that connect them to ensure proper solder flow. However, shrinking pads give you more room to snake traces between pads. For now, the pad sizes are defined in Eagle's library, but you will eventually make your own libraries. The real danger is making a pad that will disappear when you drill it. Professional board houses usually need at list 5 mils around a hole. When making a board by hand, you'll need more.

Ultimately, the pad has to be large enough to support a hole that the component's lead will pass through. You normally want the hole to be as large as the component lead plus 6 to 20 mils of extra clearance (hole plating will reduce the size of the drill somewhat). A #24 wire, for example, is 20.1 mils in diameter (see Table 3-2), so a hole to clear a #24 wire should be around 35 mils in diameter. If you want an additional 15 mils around the hole, the pad should be 65 mils (15 mils on each side means you must add 30 mils).

You can also use special shapes for pads. A round or octagonal pad makes a nice solder joint. However, it is common to use oblong pads, which give you two wide areas for solder adhesion, but still allow the maximum clearance for routing traces between the pads. Many library parts will use a differently shaped pad to indicate pin 1 of the device.

Surface mount pads, of course, don't have holes. The exact shape and size of the pad will depend on the component and the soldering method used. However, the Eagle libraries will provide suitable default values, so unless you are building custom libraries (see Chapter 5, "Custom Libraries") you won't have any problems.

Other special cases where trace width is important are outside the scope of this book. For example, very high frequency radio circuits can be tricky because traces can act like capacitors and inductors. In fact, clever designers will use this to good advantage by creating circuit elements from the PCB tracks. You can also form low-value resistors (such as the kind you might use for measuring current with values of just a few tenths of an ohm) using long traces.

1 mil = .001"

Table 3-2

Common wire sizes

AWG	Diameter (mils)	Dia (mm)
0	324.85	8.2513
1	289.29	7.3480
2	57.62	6.5436
3	29.42	5.8272
4	204.30	5.1893
5	181.94	4.6212
6	162.02	4.1153
7	144.28	3.6648
8	128.49	3.2636
9	114.42	2.9063
10	101.90	2.5881
11	90.741	2.3048
12	80.807	2.0525
13	71.961	1.8278
14	64.083	1.6277
15	57.067	1.4495
16	50.820	1.2908
17	45.257	1.1495
18	40.302	1.0237
19	35.890	0.9116
20	31.961	0.8118
21	28.462	0.7229
22	25.346	0.6438
23	22.572	0.5733
24	20.101	0.5106
25	17.900	0.4547
26	15.940	0.4049
27	14.195	0.3606
28	12.641	0.3211
29	11.257	0.2859
30	10.025	0.2546
31	8.9276	0.2268
32	7.9503	0.2019
33	7.0799	0.1798
34	6.3048	0.1601
35	5.6146	0.1426
36	5.0000	0.1270
37	4.4526	0.1131
38	3.9652	0.1007
39	3.5311	0.0897
40	3.1445	0.0799

Copper Fills

Certain types of boards—such as those containing sensitive analog amplifiers or analog-to-digital converters—can benefit by having noise-sensitive traces surrounded by ground traces. This is known as a ground ring. To carry this idea further, you can actually pour a copper ground plane over the entire board. Eagle will cut out areas for the traces. This leaves plenty of ground area surrounding the circuit. Some techniques for making boards by hand have trouble dealing with large areas of copper. Mechanical milling, on the other hand, leaves a ground plane like this, even if you don't design it that way, because the machine removes only the copper that is absolutely necessary.

To pour a ground plane, draw a polygon in the layer where you want the plane to appear. Polygons can have names, so if you want the plane to connect to, say, the GND signal, you need to name the polygon GND. Just as nets with the same name connect, polygons also connect to nets with the same name.

Polygons don't always draw correctly as you make changes. That's because placing traces and pads inside the polygon causes Eagle to carve out space for them. When you want to have Eagle recalculate (and redraw) the polygon, you can click the Ratsnest command.

Several special terms are associated with polygons. You can change these items using the Change command:

- **Pour**   Eagle can generate a solid fill pattern or a crosshatch pattern made of individual lines.

- **Width**   The polygon is actually comprised of lines. Using wide lines makes a coarse shape but generates smaller output files.

- **Rank**   When you draw a trace over a polygon, Eagle cuts the polygon out to accommodate the trace. But what happens when you draw one polygon over another? One polygon has to make way for the other. The rank determines which polygon gets out of the way. A polygon with rank 1 will override any other polygon. Polygons with rank 6 will always get out of the way of lower-numbered polygons. If two polygons with the same rank overlap, the design rule checker will decide what to do. You should set the rank to different values for any overlapping polygons.

- **Spacing**   When using a crosshatch pattern, the spacing parameter determines how close the lines that make up the pattern will be.

- **Isolate**   The polygon keeps a predetermined distance away from other elements. The Isolate parameter determines this distance.

- **Thermals**   A pad that connects to a polygon will just appear in the plane. However, if a hole is in the ground plane, it is difficult to solder the lead in place. That's because the large area of copper will draw heat away from the solder joint quickly. For this reason, Eagle places thermals around these pads. Simply put, a thermal is just a cutout in the plane around the pad. Four thin tracks automatically connect the pad to the ground plane. This improves solderability and still maintains the ground's integrity.

Figure 3-6 shows two polygons. The polygon to the left has thermals turned on (see the cross-shaped connection in the center pad of the top row of pads). The one on the right, however, has thermals turned off—the pad is simply part of the polygon.

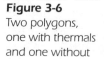

Try and Try Again

Laying out a board by hand is a lot like solving a maze. Sometimes you realize you've worked into a corner and you have no choice but to back up. That's what the Ripup command is for. When you click the Ripup tool and then click on a trace, you'll convert it back into an airwire.

Figure 3-6
Two polygons,
one with thermals
and one without

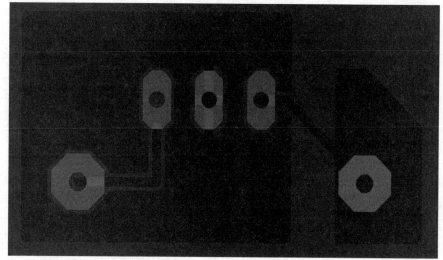

If you are really in trouble, you can click Ripup and then click Go (the traffic light icon in the toolbar). This will rip up all traces. If you just want to rip up a few signals, you can go to the command line (the box at the top of the drawing) and enter RIPUP followed by a list of net names.

Another way to rip up traces is to specify a list of traces you don't want to rip. You do this by using the Ripup command (on the command line) followed by an exclamation point and the list of signals you want to remain routed.

A common strategy is to route the power and ground busses (which should be large, short, and direct), and then you attempt to route the remaining traces (or use the autorouter as you'll see in the next chapter). If you get stuck, you might issue this command:

RIPUP ! VCC GND

This will remove everything except the power supply routing. Another trick is to click Ripup and then enter an exclamation mark into the command line. Press Enter (not the Go command). Eagle will then enable you to click on traces you want to keep. When you click Go, Eagle will rip up the remaining traces.

Checking Your Work

Eagle can check your work for you by using the DRC command. It checks that traces haven't been placed too close together, that holes are not too large for their pads, and that everything has sufficient clearance.

You can change the parameters Eagle uses as design rules and save them in files. You may want to use wider spacing, for example, if you are building boards by hand. In addition, a board maker may have particular requirements that you should check in the rule checking. In addition to testing, the autorouter (see Chapter 4) will also use some of these parameters.

The DRC dialog has nine tabs (see Figures 3-7 to 3-15). Most of these are pretty easy to understand:

- **File** The file tab lets you save and load sets of rules.

- **Clearance** You can set the minimum distance between elements on this tab. One set of measurements is used for traces that belong to different signals and another is used for signals of the same net.

- **Distance** This tab is similar to Clearance but sets the distances between elements and the board edge. It also controls the spacing allowed between drill holes.

Figure 3-7
The DRC File page

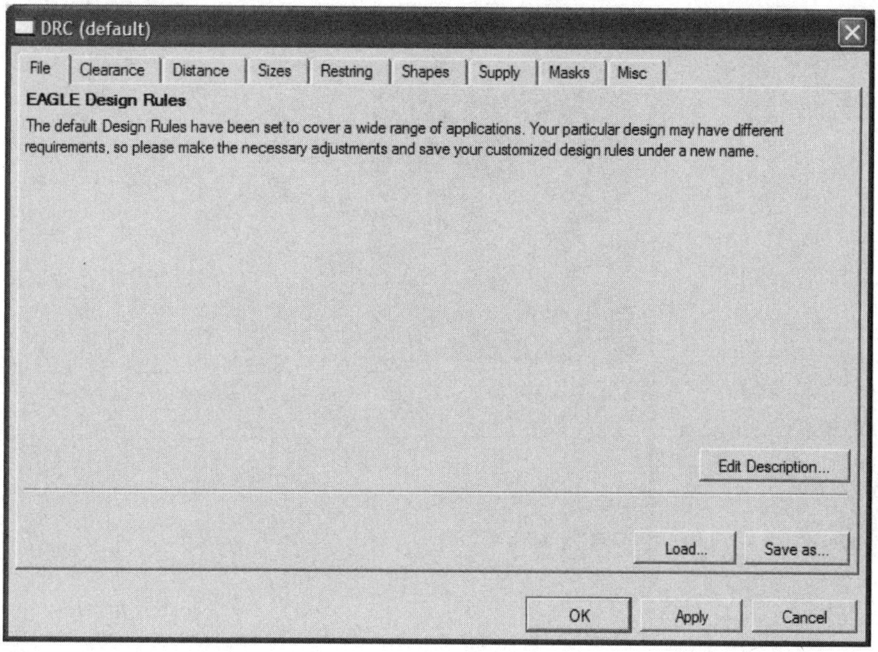

Figure 3-8
The Clearance tab

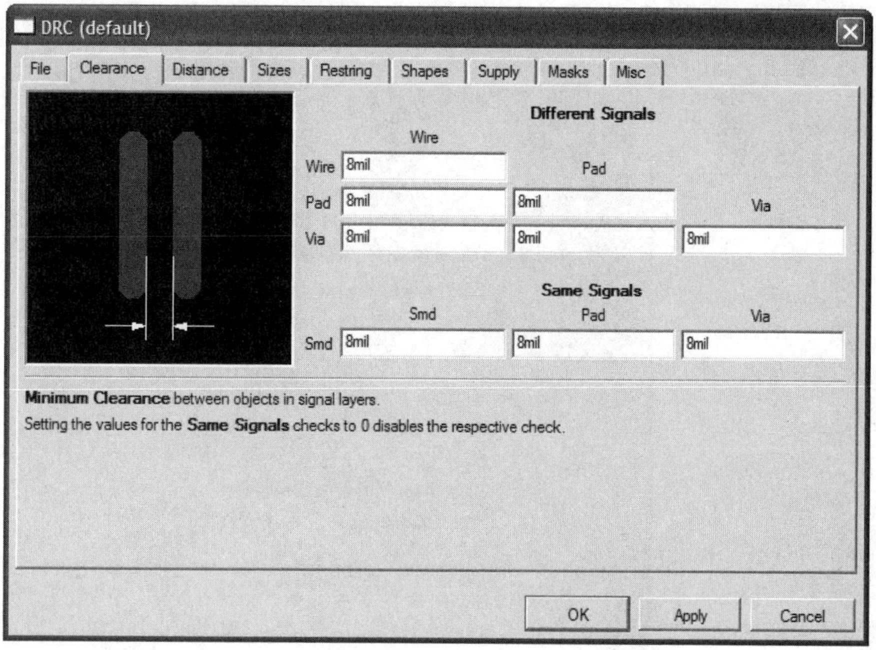

DRC (default)

File | Clearance | Distance | Sizes | Restring | Shapes | Supply | Masks | Misc

Copper/Dimension 40mil

Drill/Hole 8mil

Minimum Distance between objects in signal layers (pads, smds and any copper connected to them) and the board dimensions, and between drill holes.

OK | Apply | Cancel

DRC (default)

File | Clearance | Distance | Sizes | Restring | Shapes | Supply | Masks | Misc

Minimum Width 10mil

Minimum Drill 24mil

Minimum Sizes of objects in signal layers and of drill holes.

OK | Apply | Cancel

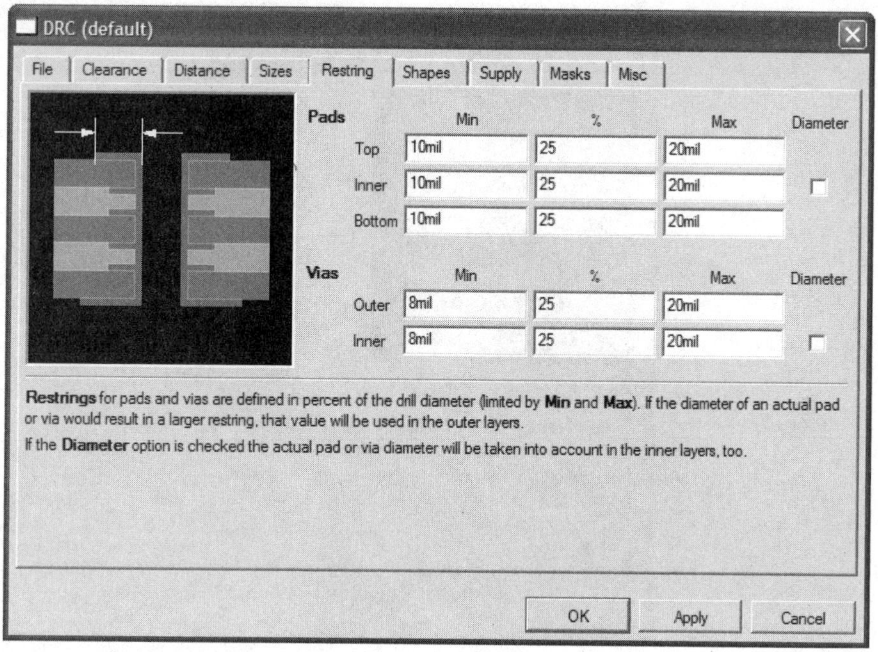

Figure 3-11
The Restring tab enables you to specify the amount of copper around holes and pads.

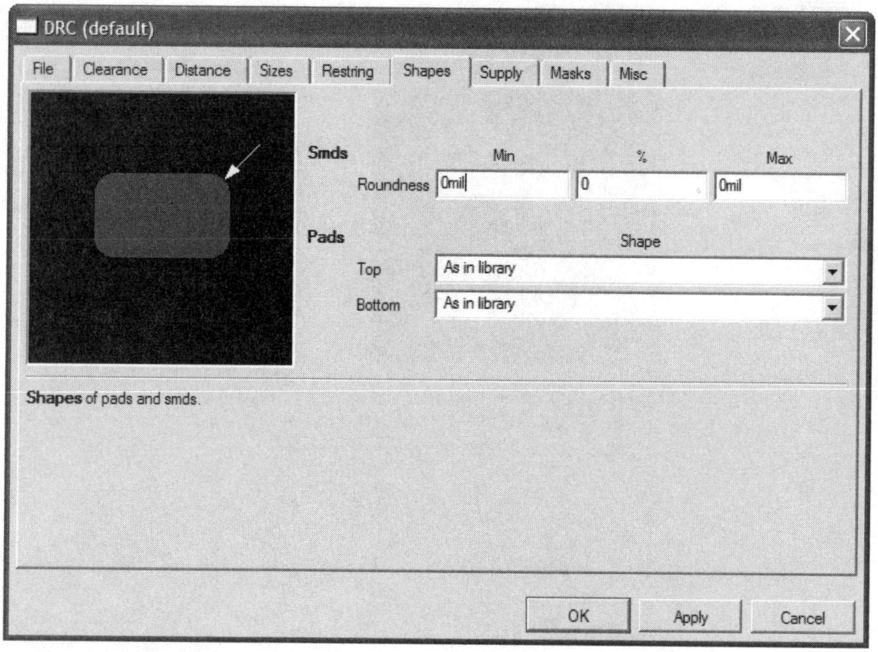

Figure 3-12
The DRC tool's Shapes tab

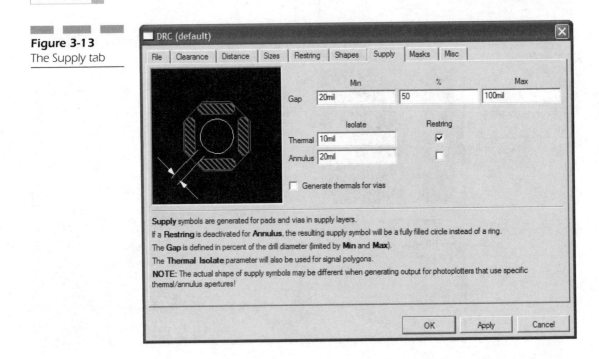

DRC (default)

File | Clearance | Distance | Sizes | Restring | Shapes | **Supply** | Masks | Misc

	Min	%	Max
Gap	20mil	50	100mil

	Isolate	Restring
Thermal	10mil	☑
Annulus	20mil	☐

☐ Generate thermals for vias

Supply symbols are generated for pads and vias in supply layers.
If a **Restring** is deactivated for **Annulus**, the resulting supply symbol will be a fully filled circle instead of a ring.
The **Gap** is defined in percent of the drill diameter (limited by **Min** and **Max**).
The **Thermal Isolate** parameter will also be used for signal polygons.
NOTE: The actual shape of supply symbols may be different when generating output for photoplotters that use specific thermal/annulus apertures!

OK | Apply | Cancel

DRC (default)

File | Clearance | Distance | Sizes | Restring | Shapes | Supply | **Masks** | Misc

	Min	%	Max
Stop	4mil	100	4mil
Cream	0mil	0	0mil
Limit	0mil		

Mask values are defined in percent of the smaller dimension of smds, pads and vias (limited by **Min** and **Max**).
Stop masks are generated for smds, pads and those vias that have a drill diameter that exceeds **Limit**.
Cream masks are generated for smds only.

OK | Apply | Cancel

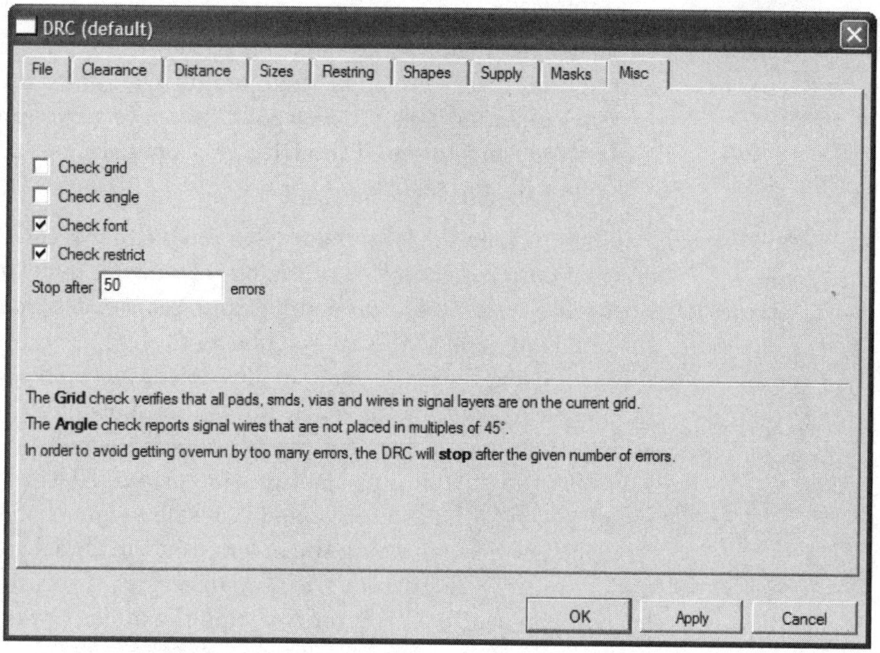

Figure 3-15
You can set
several options on
the Misc tab of the
DRC tool.

- **Sizes**   You can set the smallest allowed trace and hole sizes on this tab.

- **Restring**   The Restring tab sets the amount of copper required to remain around a drill hole (a pad or via).

- **Shapes**   This tab can change the shape of *Surface Mount Device* (SMD) pads and also of regular pads. Usually, you want to use the pad shape specified in the Eagle libraries, but it is possible to override the shapes from this dialog. When you press Apply, Eagle will adjust the pads per your request. If you change the pad shapes, you can also return the setting to As In Library to restore the shapes to the default.

- **Supply**   The Supply tab is primarily of interest when you are working with boards that have interior layers.

- **Masks**   Eagle can generate solder mask and solder cream layers. These layers are typically somewhat larger than the pad over which they are located. This page sets the minimum and maximum percentages to which the masks are allowed to expand their size.

■ **Misc**   As its name implies, this tab has a few miscellaneous tasks. In particular, it can test to make sure everything lines up on the current grid, that traces are on 45-degree multiples, and that nothing is in a restricted area. This tab also enables you to set a number of errors that you want to see. If the DRC check generates this many errors, Eagle will stop testing.

When you run the DRC (click OK), you'll either receive a message that no errors occurred, or you'll get a dialog showing all the errors. The error locations will also have special highlighting. You can double-click on an error in the dialog to jump to the area in question.

When using libraries, you may have clearance and other errors that you didn't cause. Don't be too quick to assume that a DRC error is absolutely wrong. However, be very certain before you dismiss an error. Many times Eagle is trying to tell you something important. However, many cases occur where you get a false positive and it is safe to ignore the error.

If you get tired of seeing the errors, you can click Del (or Del All) on the dialog to delete the current error (or all errors). If you dismiss the error dialog, you can use Tools | Errors to bring the dialog back.

Final Checks

Once you get a clean DRC and the Ratsnest command reports 0 airwires, you should carefully examine your handiwork. Be certain that you have any mounting holes and silk-screen markings that you need. In particular, if you have polarized components (like electrolytic capacitors), be sure the plus markings are visible. Any significant marks (like pin 1 of connectors) should be visible after the parts are installed.

At the very end, be sure to make one last mock-up board and put parts in it. It is easy to get excited and want to start making the board (or send the files off for production), but as annoying as it is to wait, it is much cheaper to fix problems now than after you've etched a board.

One other thing to watch for is an excessive number of vias. Each via is a hole to drill (and therefore an expense). Each via is also a possible point of failure. Most boards will have vias, but you should always try to minimize the number of vias on any board.

For Practice

Practice makes perfect. However, you won't want to start practicing on a giant, complicated board. Start with some simple projects. For example, the power supply in Figure 3-16 is a good starter project (the schematic is on the included CD-ROM). Try routing it as a two-sided board. Then try again using only the bottom layer (a single-sided board).

If you really want a challenge, lay out the board and then try reducing the board size and lay it out again. You'll eventually discover how small you can make the board before you have no chance of routing it.

In extreme cases, you can also install wire jumpers to give yourself an extra pseudolayer. However, you'll have to add the jumpers into the schematic (using, for example, wire pads) so that the freeware Eagle program will recognize that they don't violate layout constraints.

Figure 3-16
Practice laying out this power supply.

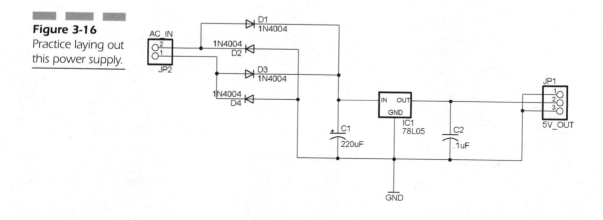

In Summary

If you enjoy solving puzzles, you'll probably like routing PCBs. If you are like me, you'll probably hate it! The good news is you will get better with practice. With Eagle you can keep attempting different layouts until you are satisfied.

In the next chapter, you'll read about the autorouter that can lay out a board automatically. That sounds ideal. However, if the autorouter will only route 97 percent of your board, it might as well route 0 percent of it. Don't get me wrong, I like the autorouter, but it is not a magic wand—it is just another tool.

CHAPTER 4

Autorouting

I've worked for quite a few big companies over the years, and I have found one characteristic they all have in common: monkey money. Monkey money is a term I coined to describe how you get a larger budget at a big company. Asking for more staff, for example, is likely to be a waste of time. No budget exists for something like extra employees. However, you can ask for—and receive—much more money if you ask for monkey money. Monkey money is money spent researching a magic machine or process that will allow everyone's job to be performed by a monkey.

My experience is that it is safe to ask for monkey money because in real life monkeys can't do very much, but management is always happy to write a check on the promise that they will be able to fire everyone and replace their workers with simian labor.

Autorouters are somewhat like monkey money projects. In theory, you can draw a schematic, form a board outline, place the components, and press the autoroute command to finish the board. As you might expect, things aren't really this simple. Eagle's autorouter is quite good and for many boards it will do the job. However, you might not find the autorouter very useful in several situations, such as the following examples:

- Sometimes the autorouter can only route most of the board (perhaps 98 percent). The two or three remaining airwires will be impossible to hand route without making changes.
- The autorouter sometimes makes inefficient routes or uses a large number of vias.
- You may wish to control certain critical traces or traces you are using to form components (such as shunt resistors or inductors).

These situations are often correctable. For example, increasing the board's size may allow the autorouter to complete its task. Rearranging a few components can often have a dramatic effect on a board's routability. You can also modify parameters in the design rules to affect Eagle's autorouter.

If you have special routing requirements, it is possible to manually route a portion of the board and then invoke the autorouter to try to complete the remaining traces. In fact, this is often a good strategy. I often route the power and ground busses along with any high-speed clock signals by hand. Then I'll let the autorouter do the remaining layout.

Don't get me wrong. The autorouter is a great tool, but it won't allow a monkey to lay out a board anytime soon.

About Autorouting

The Eagle autorouter is a rip up/retry router. That means it starts placing tracks until it works itself into a dead end. Then it rips up some tracks and tries a different arrangement. You can control how many tracks the autorouter can rip up. The more tracks you allow, the more likely the autorouter will find a solution for your board, but the longer the process will take.

Eagle claims that the autorouter can route any board that is not physically unroutable. However, it might take an infinite amount of time to route any particular board completely. The autorouter uses information from the design rules, the net classes, and an autorouter control file to decide what to do. The control file lets you set things such as the number of layers, the directions preferred on each layer, and which autorouting steps to perform.

The autorouter uses three steps:

1. **Bus routing**   The autorouter considers busses as any connection that can be realized as a straight (or nearly straight) line.
2. **Main routing**   The main routing step allows a large number of vias and connects any routes not already connected.
3. **Optimize**   The autorouter routing can perform multiple optimization steps after it's complete. Each pass attempts to reroute each track using fewer vias and lines that have fewer bends.

You'll usually use the first two steps along with multiple optimize steps. However, you can turn off each step and control how many optimize passes the router makes.

When Things Go Right

Ideally, Eagle's default autorouting rules will do a pretty fair job provided your board is not too dense or complex. As an example, consider Figure 4-1. This schematic shows a simple *light-emitting diode* (LED) display that uses a Microchip PIC12F629 microcontroller. In addition to the LEDs and the processor, a simple 5V power supply (using a 78L05) is also on board. You could easily power the circuit with a 9V battery (attached to the two wire pads).

Figure 4-1

A PIC-based LED
blinker

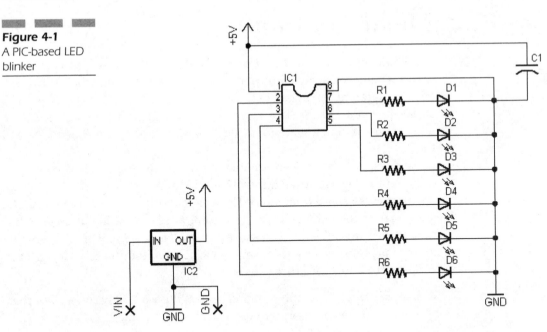

Figure 4-2 shows one possible layout for the board's components, and the board, as shown, is ready to route. The board uses two different net classes (see Figure 4-3). The default net class is set to use 10 mil traces and 10 mil spacing. The power class is used for the input voltage to the regulator, the 5V supply for the microprocessor, and the ground traces.

This board has enough open area and few enough parts that the autorouter should have no problem with it. If you press the Autoroute toolbar button, you'll see a dialog like the one in Figure 4-4. For now, ignore all the tabs except for the first one. You'll see two combo boxes, one for the top layer and another for the bottom layer. These select the preferred direction of traces for each layer. With the top layer set to vertical (|) and the bottom layer set to horizontal (-), the autorouter produces the board seen in Figure 4-5. Notice that the power conductors are wider than the other traces (obeying the net class specifications).

In this case, when the autorouter completes, you'll see a message in the status bar that indicates the board is 100 percent routed. That's good news. Sometimes you will see a number less than 100 percent, which means you will have more work to do. You can also hit the Ratsnest button and see that it reports zero airwires.

If you don't like the layout of the board, you can always pick the Rip up tool and then click Go. This will remove all the traces and you can rerun the

Figure 4-2
One layout for the
LED blinker

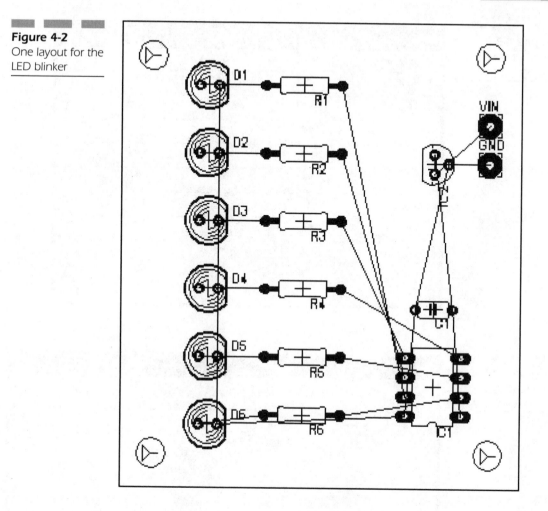

autorouter. For example, Figure 4-6 shows the same board routed with the
bottom layer vertical and the top layer horizontal.

You may not relish ripping up the board if you manually laid out some
tracks. Remember, you can issue the RIPUP command from the command
line and exclude certain tracks. For example, if you've manually routed the
VCC, GND, and CLK traces, you might issue the command:

RIPUP ! VCC GND CLK

It is usually best to set one layer to prefer one direction and the other
layer to prefer the complimentary direction. You can select * for either layer,
which makes the autorouter have no direction preference for that layer.

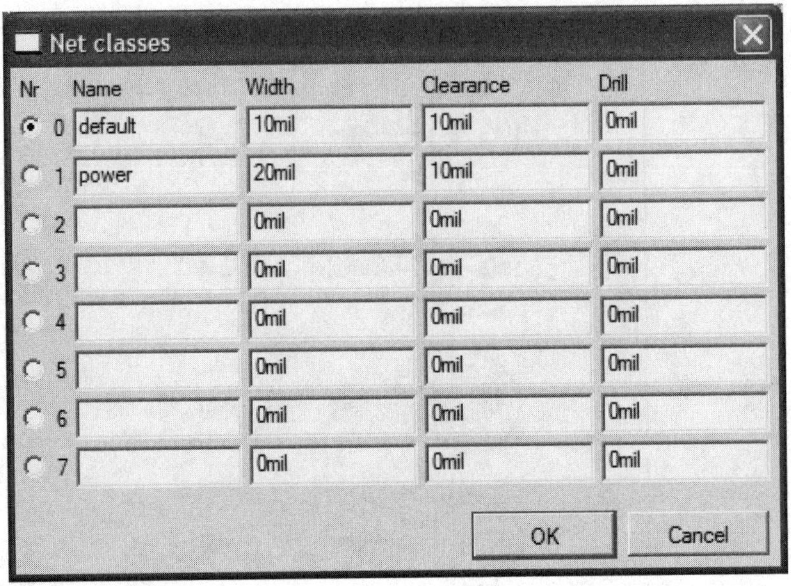

Figure 4-3
The Net classes
dialog

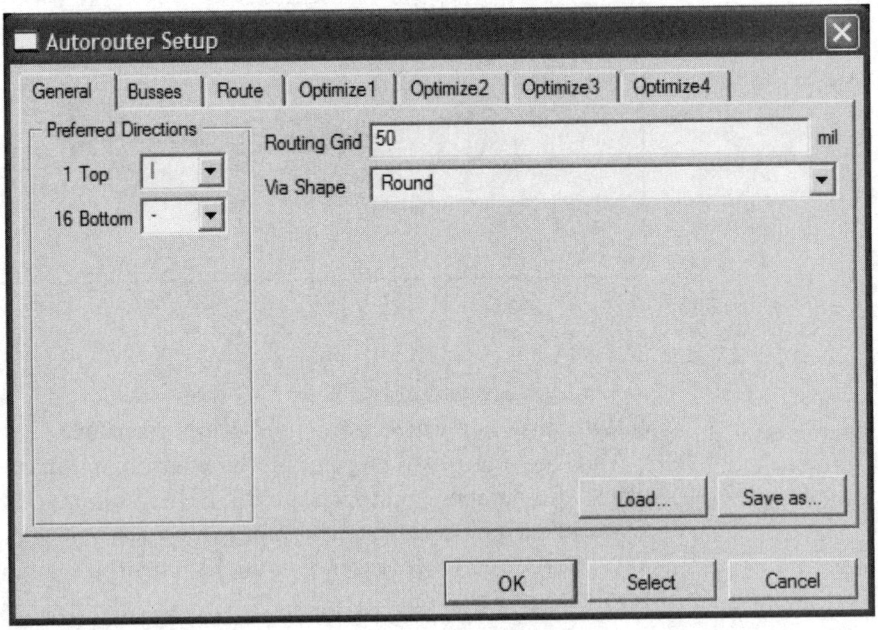

Net classes

Nr	Name	Width	Clearance	Drill
⊙ 0	default	10mil	10mil	0mil
○ 1	power	20mil	10mil	0mil
○ 2		0mil	0mil	0mil
○ 3		0mil	0mil	0mil
○ 4		0mil	0mil	0mil
○ 5		0mil	0mil	0mil
○ 6		0mil	0mil	0mil
○ 7		0mil	0mil	0mil

OK   Cancel

Figure 4-4
Starting the
autorouter

Autorouter Setup

General | Busses | Route | Optimize1 | Optimize2 | Optimize3 | Optimize4

Preferred Directions

1 Top [|]

16 Bottom [-]

Routing Grid 50 mil

Via Shape Round

Load...   Save as...

OK   Select   Cancel

Figure 4-5
The blinker
autorouted (top
layer vertical)

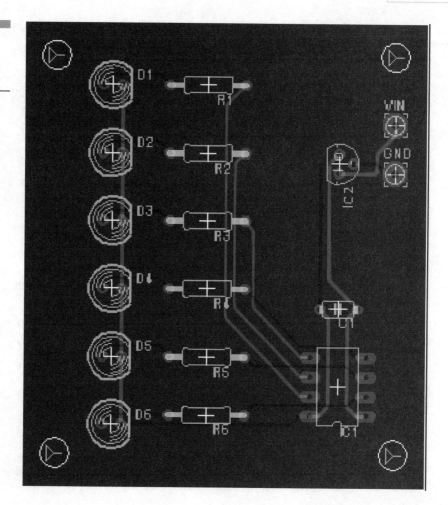

However, it is usually more efficient to try to keep traces on one side of the board going in one direction or the other. The exceptions are when you are routing boards with surface-mount components or single-sided boards. You should use the * selection for these types of boards.

This board is wide open enough that it does not require any vias (the router tries to join the top and bottom at component pins where possible).

Figure 4-6
The blinker
autorouted
(bottom layer
vertical)

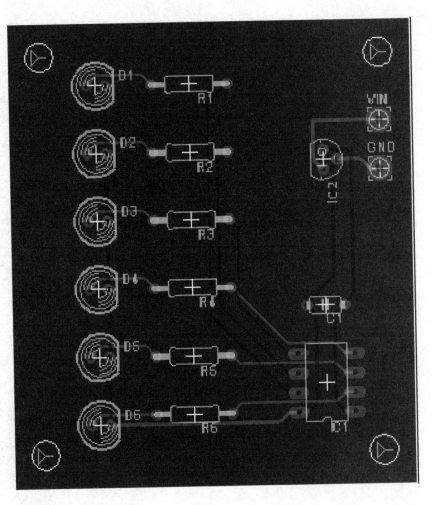

Fine-Tuning

You may want to rip up the autorouting and change other parameters. For example, if you are building boards by hand, you may want to make all the traces 25 mils in width. If you do this, you'll find the board can't be routed on one side by the default autorouter parameters (see Figure 4-7). The router can't find a path for two traces.

Sometimes you can experiment with other arrangements of components and the autorouter will be able to find a path. With a little practice, you'll be able to recognize where the leftover airwires are and guess what

Figure 4-7
With thick traces
on one side, the
board is
unroutable in this
configuration.

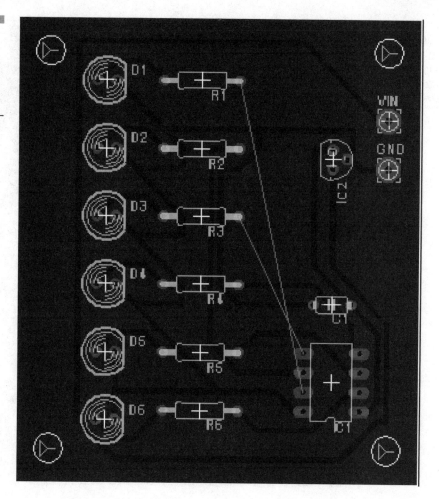

components you can move for the greatest effect. Figure 4-8 shows a slightly different arrangement that will successfully route with 25 mil traces and spacing.

Referring back to Figure 4-7, you can see that the airwires remaining stretch from the microcontroller near the bottom of the board to the LEDs near the top. It makes sense that relocating the *integrated circuit* (IC) to a more central location would ease routing. However, putting the IC in the center congests the area that contains the power supply circuit. That's why the power supply is split in Figure 4-8. The regulator is below the microprocessor and the wire pads for voltage input are above the microcontroller.

Figure 4-8
With slight
adjustments, the
board can be
routed.

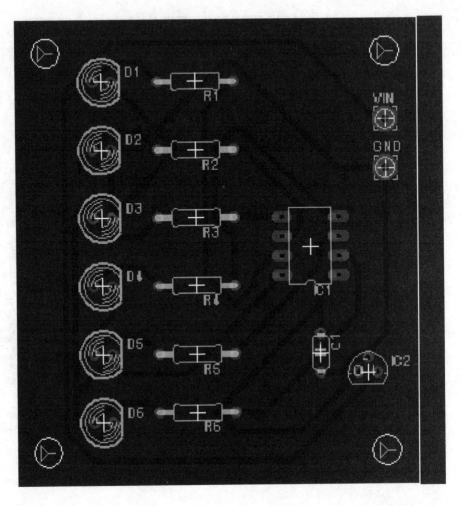

This causes some routing to occur at the far right-hand side of the board, but that area is not used otherwise, so that's not much of a concern.

This example is, of course, ideal. In real life, you may not have so much area on the board. Of course, a double-sided board has twice as much area, but these are difficult to make by hand. If you plan to manufacture your own boards, you may want to stick with single-sided boards with wide traces and just allow a little more area on the board.

Grids

You've already seen that net classes can alter the operation of the autorouter. Any parameter not set in the net class dialog will take on a value defined by the design rules. Another factor to consider is the placement grid that the components use. The placement grid is involved with autorouting because the autorouter also uses a grid and these two separate grids interact.

For example, suppose you have some tiny surface-mount components that have leads spaced at .05 inches; the routing grid must be .05 inches (50 mils) or less. That means the placement grid should also be 50 mils. If it were less, you might put a component down in such a way that the autorouter could not connect to it.

You can make the placement grid larger than the routing grid, but the placement grid must be an integer multiple of the autorouting grid. This is common with through-hole designs. You probably want to place components on a 0.1-inch grid (the spacing for a common IC). However, you want the autorouter to work on a 0.05-inch grid so it can run thin traces between IC leads. However, if the placement grid were, for example, 0.1 inches and the autorouter grid were, for example, 0.04 inches, the router might not be able to connect to all the pins.

This can be tricky when using libraries from different sources. ICs, of course, have a required spacing, but other components can potentially be laid out on a variety of grid spacings. So, if some parts are on a 0.1-inch grid and others are on a 2-millimeter grid, you might have a problem. Luckily, if you stick with the Eagle standard libraries, you shouldn't have problems. Besides, the components are clearly marked as to their design grid. For example, the 0207/12 resistor is clearly marked as type 0207 and a 12-millimeter grid.

Some surface-mount components may have to reside off the grid. In these cases, each pad must be large enough so that it covers at least one routing grid. Otherwise, the router may be unable to connect to the pad and will issue an error message. You can set the routing grid on the General tab of the autorouter command dialog.

With all this concern over the grid, you might be tempted to just set a fine routing grid size (maybe 5 mils) and forget it. That will work, but remember that the more grid lines involved in routing, the more memory the route process will require. It also increases the routing time, as you would expect.

Working with Layers

You've already seen that you can select preferred directions for boards. You can also disable a layer (for example, for single-sided boards you disable the top layer). The free version of Eagle only supports two sides, but if you have a different license you can route multiple layers. If you are using inner layers, be sure to pick them in pairs along the outside. For example, you'd use layers 2 and 15 first. Then, if you need more layers, you'd pick 3 and 14.

When routing a board with through-hole components, the direction for each layer is not usually very important. You'll just make sure that the top layer and bottom layer are 90 degrees from each other. With surface mount, you can often see from the airwires that one direction will have an advantage over another on a given side. However, for many surface-mount boards, it is better to select * to allow the autorouter to freely arrange the traces.

Several other layers influence the autorouter. First, the dimension layer —where you normally draw your board's outline—acts as a boundary for the autorouter. The router won't draw traces outside the outline in the dimension layer.

You can also place rectangles, polygons, and circles in any of three restrict layers to prevent the router from placing tracks in a given area. The tRestrict layer covers the top layer. The bottom layer uses the bRestrict layer, and the vRestrict layer prevents vias and holes in the specified area. These are useful if you have, for example, a heat sink that will mount flat against the board. Many components in the standard library define restricted areas where necessary.

You can also draw polygons in the dimension layer to restrict all signals from an area. However, this can be confusing when you have the board made because this layer usually defines the perimeter of the board. If you aren't careful, you may get a cutout in the middle of your board corresponding to a polygon in the dimension layer!

Backups

As the autorouter does its work, it periodically saves temporary results to a .job file. If the computer crashes or you stop routing (by pressing the stop sign or Control+Break) you can restart the autorouter without repeating all the previous work. Of course, the job will continue with the settings that were in force when you started the job. Any changes you make will be discarded unless you tell Eagle to discard the old job and start over.

The backup files are created roughly every 10 minutes. In the worst case, the router may lose up to 10 minutes of work in the event of an interruption.

Controlling Routing

When you start the autorouter, it displays a dialog with several tabs. You can use these tabs to control the operation of the router. You've already seen how you can set the routing grid and preferred directions. The other choice on this dialog is the default via shape to use (round or octagon). Also, buttons on this page enable you to save (and load) your settings so you can apply the same parameters to different autorouting runs.

The second and third tabs are nearly identical. They set parameters used by the bus router and the main router, respectively. Each box on the dialog enables you to set a cost or a maximum allowed count for certain operations. For example, the Bus router has NonPref set to 4 by default. That means that placing a track in a nonpreferred direction has a cost of 4. Routes with lower costs are preferred by the router.

In addition to the layer costs, you can set several other variables:

- **Via**   The cost of using a via.

- **NonPref**   Placing tracks in a different direction than the preferred direction. If this is set to 99, the router will *never* route against the preferred direction.

- **ChangeDir**   This parameter sets the cost of bending a trace.

- **OrthStep, DiagStep**   These set the router's preference for making corners versus making diagonal traces. Eagle notes you should not alter these parameters lightly.

- **ExtedStep**   The cost of making traces at a 45-degree angle to the preferred direction. Obviously, this only applies to layers that have a preferred direction.

- **PadImpact, SmdImpact**   Pads and *surface mount* (SMD) pads produce good and bad areas that the router prefers to fill (good areas) or prefers to avoid (bad areas). The higher these values, the further traces will go toward the pads.

- **BonusStep, MalusStep**   These costs determine how much difference is afforded good (Bonus) and bad (Malus) areas (see PadImpact and SmdImpact).

- **BusImpact**   For the bus router, this cost determines how direct bus lines are routed.

- **Hugging**   Higher values in this parameter cause parallel traces to be closely spaced.

- **Avoid**   When the router rips up a track, it tries to avoid the area it was in originally. Higher costs in this parameter cause the router to increase its avoidance of this area.

- **Polygon**   The router can place tracks through a polygon (copper fill). This has the effect of breaking the copper fill into pieces. Higher costs in this parameter make it more likely that the router will go around a copper fill area.

You can also set several counts:

- **Via**   This count sets the maximum number of vias that can be used for one trace.

- **Segments**   You can control the maximum number of segments used to create one track.

- **ExtdSteps**   This count controls how many 45-degree steps away from the preferred direction are allowed before invoking the ExtdStep cost.

- **RipupLevels**   The number of tracks that can be ripped up to break a dead end while routing.

- **RipupSteps**   When a track is ripped up and rerouted, it may trigger another ripup step. This count controls the maximum number of times this may occur.

- **RipupTotal**   The total number of tracks that can be ripped up at once. This number is a guideline and may be exceeded in some cases.

Not all of these parameters apply to each routing phase. For example, the ripup counts only apply to the main router. Also, some steps normally don't allow certain operations. For example, the bus router has the via count set to zero, which effectively makes the via cost irrelevant.

Each page also has an active check box. You can disable any step you like by unchecking the active box. You probably don't want to disable any of the steps. Also, Eagle recommends you try not to change the autorouter parameters unless you are certain you need to do so.

Optimization

Once the autorouter finishes with the bus routing and main routing steps, you can run one or more optimizer steps. This is usually done because the normal router will use quite a few vias and other unnecessary items. The optimizer steps are similar to routing (and share many of the same parameters). However, the parameters for optimization are typically more strict than the routing parameters. The goal in routing is just to find a possible route. The optimization passes aim to reduce bends and vias.

The Route tab has an Add button that adds extra optimize steps. In addition, each Optimize tab has an Add button for adding a new step and a Del button for removing the current tab. You can also selectively disable optimize steps by using the Active check box just as you can disable any of the tabs.

Single-Sided Routing Tips

If you want to route a single-sided board, you'll usually set the top layer to N/A in the General tab, and the autorouter will dutifully avoid putting traces on the top of the board. However, you may have to enlarge your board to make everything fit. In many cases, you'd be better off making a smaller board and using jumpers to connect the traces that can't be routed.

This can be accomplished in several ways. If you have a good idea of where the problem lies (perhaps from failed attempts at routing), you can install 0-ohm resistors in the critical paths. The router will treat these as full-fledged components, but when you assemble the board you can simply use wire. This is so common the standard Eagle library has a section (called Jumper) that contains various jumper patterns.

Another approach is to route with two sides, but set the cost of the top side to a very high value. The only problem with this approach is that the router may make some connections to component pins, or place tracks under components, which is not very convenient when it is time to assemble the boards. To prevent these problems you can draw restriction rectangles over the components in the tRestrict layer. This will prevent top traces from connecting directly to components or crossing under components.

Figure 4-9 shows the LED flasher board routed in this manner. Notice the rectangles covering each component. Without them, traces would pass under components and might connect to component leads.

Figure 4-9
Single-sided
routing

When it is time to make the board, you only make the bottom layer. You then insert wires into the vias and use the wires to complete the top side of the board. Because of this, you may want to change the vias to be larger than normal. Eagle provides design and autorouter control files, especially for single-sided routing.

About Control Files

If you make many changes to the control files for the autorouter and the design rules, you may want a more efficient way to make modifications. If you open the files with an ordinary text editor, you'll see that they have a simple file format.

A text editor allows you to make mass changes (such as changing all occurrences of 8 mils to 12 mils). This can be easier than working through the graphical user interface.

However, you probably shouldn't alter the control files in an editor or in the Eagle interface unless you are certain you know what you are doing. Changes to the autorouter control file, in particular, can easily do more harm than good.

Coaxing the Autorouter

Sometimes you get lucky and your board routes 100 percent the first time. You can try several things to help the autorouter:

- Rearrange the components.
- Enlarge the board's dimensions.
- Reduce the trace width.
- Reduce the trace spacing.
- Reduce the routing grid size.
- Increase the layer count.
- Attempt to route problem areas by hand before starting the autorouter.

Routing—even automatic routing—is like working a puzzle or solving a maze. When you find a dead end, you have to back up and try again. It is usually helpful to look at the pattern of airwires after routing to see where the problems occur. Attacking these problem areas will usually at least get you closer to a fully routed board.

About This Project

The circuit board in this chapter uses a tiny 8-pin Microchip PIC microcontroller. With the right configuration, the PIC12F629 used can drive 5 LEDs (this board has an unused space for an LED on pin 4 which is only an input on this particular PIC). In this configuration, the chip uses an onboard clock and reset circuit, so practically every pin is available for use.

The CDROM includes the parts list and a hex file that must be programmed on the PIC to make the LEDs blink in a pattern. Of course, by changing the program in the chip you can alter the behavior of the LEDs

In Summary

The autorouter is a great tool. If your boards are not too complex, the autorouter will often find an acceptable route quickly. If it doesn't, you'll have to take steps to solve any problems that prevent a solution. Often, you'll route part of the board and let the autorouter do the rest.

If your board will route with the default parameters, you are in luck. If the default parameters won't work, you'll have to delve into the deep meanings of all the autorouter control file parameters. Some of these are not very intuitive, but with the information in this chapter and the Eagle documentation, you can exercise great control—maybe too much control—over the autorouter.

At this point, you can produce some very useful boards with Eagle. Creating schematics, autorouting, and manual routing are the three major tasks required to make boards. However, up to now, you've been dependent on the predefined Eagle libraries.

Sure, you can often use generic *Dual Inline Package* (DIP) outlines for most ICs. You can even arrange wire pads and other components to fit specialized components in a pinch. However, for the best overall results, you'll want to create your own libraries.

CHAPTER **5**

Custom
Libraries

I've written a lot of software in my life. I remember back when object-oriented programming was just appearing in the marketplace. A company that sold Smalltalk (an object-oriented language somewhat popular at the time) ran a full-page ad in several magazines that showed a Windows text editor. Inside the editor was a message that said (more or less):

This text editor was written in one line of Smalltalk code!

My boss at the time came into my office brandishing a magazine open to this ad and waved it in my face. The ad had convinced him that because Smalltalk was object-oriented, that made it easy to write text editors. The truth, of course, is that many more lines of code existed in that text editor, but they were wrapped up in the internal Smalltalk library (and, in a deeper sense, inside the Windows edit control).

A better message would have been the following:

This text editor requires 10,000 lines of code, but you can call it with one line of Smalltalk code!

Truth in advertising aside, the same situation exists in Eagle and many other *Computer-Aided Design* (CAD) programs. Sure, it's difficult to keep track of schematic symbols, package footprints, pads, keep-out areas, and the sizes of various elements. However, if you have a properly designed library of parts, much of this is invisible (or nearly invisible) to you. You simply place the parts on the board and all the various elements—pads, outlines, and restriction polygons—are placed in the correct place and in the correct proportions.

One of Eagle's strengths is that it has a very extensive library of components. You can lay out many useful boards by using the built-in components. This is especially true if you use generic parts to stand in for other parts. For example, the board in the last chapter used a generic 8-pin *Dual Inline Package* (DIP) footprint instead of one specifically for the PIC microcontroller. You can carry this idea further, if necessary. For example, suppose you have to mount a switch on the *printed circuit board* (PCB) and you don't have a library part that matches the switch's dimensions. You could place wire pads in the correct places to handle the switch. Of course, this makes your schematic ugly, but it does allow you to quickly place pads in any pattern you need. You can also find more component libraries (some supplied by users) on the Eagle web site.

However, for parts you use frequently, it is better to build a custom library to handle the part. Eagle makes this relatively easy. However, to make a well-designed library requires a good understanding of Eagle and a bit of forethought.

Inside Libraries

Each part in an Eagle library actually consists of multiple pieces bound together. In particular, the library contains a schematic symbol, the package (for example, a TO-92 or a 16-pin DIP), and the component definition. A component definition (known as a device) links a schematic symbol to a package. It also defines things like the naming conventions used, the variants available, and the exact correspondence between the logical pins of the schematic symbol and the pads on the package.

Certain exceptions don't have these three distinct items. For example, supply symbols (like Vcc or Vss) have symbols, but no packages. Frames and other decorative "components" don't have packages or pins. You can also make a library containing nothing but packages,which are useful when laying out boards without using schematic capture.

The easiest way to get a feel for this is to open an existing library using the library editor. You can open and edit a library in several ways. One way is to go to the Eagle Control Panel and select the library branch of the main tree. Find rcl.lbr (see Figure 5-1) and you'll see that the right-hand pane has a very comprehensive description of the library's contents.

Figure 5-1
Selecting a library from the Control Panel

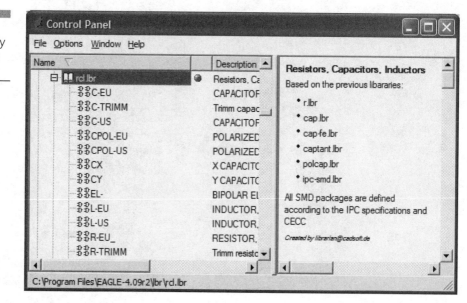

You can right-click the rcl.lbr entry and select Open to edit the library. At first, you may think you've made a mistake. The editor screen is practically blank. That's only because you start out with no particular component selected. Just to have something interesting to look at (see Figure 5-2), use the Library | Device menu command and select C-US from the resulting dialog box. Now you can see the schematic symbol (to the left), the package footprint (at the top), and a list of package and variant types. If you click the different package types, you'll see the corresponding package footprint in the top windowpane.

You can use the Library | Symbol command if you want to edit the schematic symbol. The Library | Package command edits a particular package. When you are editing a symbol or package, the toolbars look similar to when you are editing a schematic or a board (although a few subtle differences are apparent).

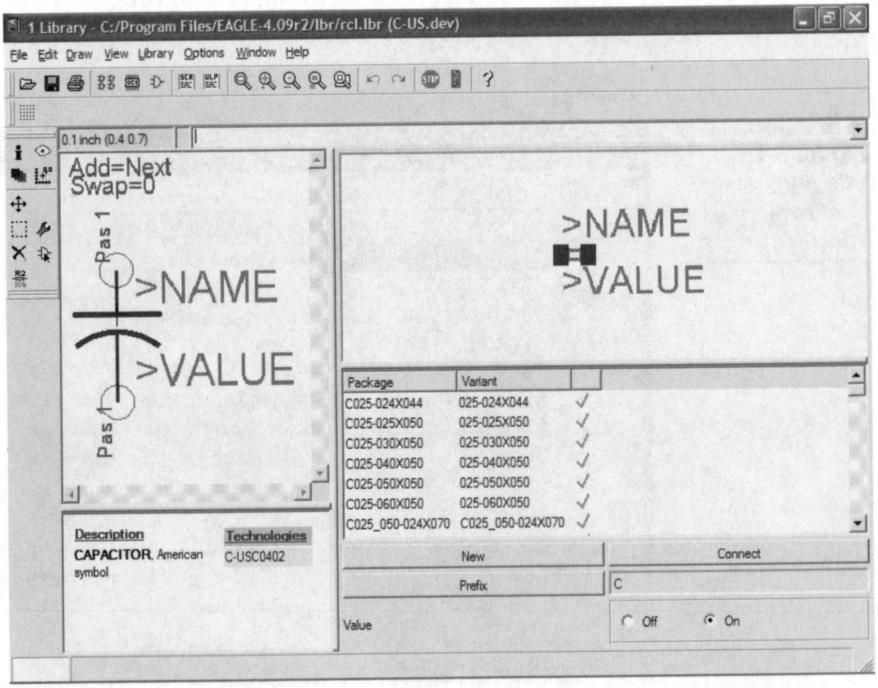

Figure 5-2
An existing component

Changing Components

One of the easiest ways to make a new component is to start with an exist- ✗
ing component that is similar. Of course, if necessary, you can create one
from scratch. To start, let's look at the steps required to make a custom
8-pin DIP.

First, start with the dil.lbr library file. This file contains a variety of
generic DIP components. Here are the first steps required:

1. Select Library | Symbol and pick DIL-8.

2. Click the Layers tool and make sure that every layer is selected to
   display.

3. Use the group command to select all items in the symbol.

4. Cut the group. You may want to use the command line so you can place
   the reference point exactly. If you want to use the command line, enter:
   CUT (0 0) ;.

5. Use File | New to create a new library.

6. Use the Library | Symbol command. From the resulting dialog, type in
   a new symbol name (for example, TEST8) and click OK. When
   prompted, agree to let Eagle create a new symbol.

7. Paste the symbol into the new workspace. For better control, you may
   want to enter the following into the command line: **PASTE (0 0) ;**.

The symbol will appear slightly different because no connections exist
between the symbol and a package, but all the elements of the symbol itself
should be present. At this point you could change the symbol however you
see fit.

Editing Symbols

Usually, everything you draw as part of the schematic should be on the
Symbols layer. You can place the text >NAME on the tNames layer to set
where the name should appear on the schematic. You can also set >VALUE
on the tValues layer to set the default position for the value set in the
schematic editor. These placeholders also set the font and text size used for
these labels.

Two major differences exist between the symbol editor and the normal schematic editor. First, an extra tool is located at the bottom of the vertical toolbar for defining pins (the parts of the symbol that connects to nets). The other is that the Change command has several different options, most of which are related to pins.

When you add a pin, the top toolbar offers several options. You can rotate the pin (just like right-clicking while placing the pin) and change its length. When drawing logic gates, you may want to show an inverting circle on the symbol (Eagle calls this a dot), or a clock symbol. Tools are available for this as well. Figure 5-3 shows a flip-flop, for example, that has these special pin decorations. The CLK pin has the clock decoration, while the PRE and CLR pins have dots indicating they are inverted.

You can also control if the name of the pin or pad is displayed (you can display both) and the pin's function (see Table 5-1).

An additional input item on the top toolbar allows you to set the swap level for the pin. This is simply a number from 0 to 255. Usually, you'll set each pin's swap level to 0 to indicate that it is a unique pin and can't be swapped with another pin. If the number is not 0, then the pin can be swapped with any other pin on the same symbol that has the same swap level.

For example, a resistor—a device with interchangeable pins—would define its pins with a swap level equal to 1. An electrolytic capacitor, on the other hand, would define both of its pins with a swap level set to 0 since the pins are not interchangeable. Of course, some devices (an AND gate, for example) might have some pins that are interchangeable and some that aren't, so one symbol may have some pins with a 0 swap level and some with a nonzero swap level.

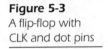

Figure 5-3
A flip-flop with
CLK and dot pins

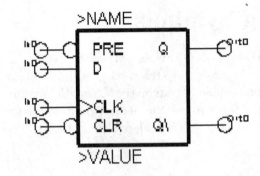

Table 5-1

Pin functions

Function	Description
NC	Not connected
In	Input
Out	Output (totem pole)
I/O	In/output (bidirectional)
OC	Open collector or open drain
Hiz	High impedance output (such as three-state)
Pas	Passive (for resistors, capacitors, and others)
Pwr	Power input pin (Vcc, Gnd, Vss, Vdd, and so on)
Sup	General supply pin (such as for a ground symbol)

The Change command enables you to alter the usual things. It also has special commands to change the pin properties (such as the length, the swap level, and so forth). You can set the visibility of pin labels as well.

Editing Packages

When you are satisfied with your symbol, you should save your work. Then you can follow the same strategy to copy an existing package into your new library. Again, the tools are mostly familiar but with a few changes. In particular, you can add pads, which are the through-hole contacts for conventional components, and SMD pads, which are for surface-mount components.

A well-drawn component will have silk-screen information drawn in the tPlace layer. However, since this layer will generate silk-screen information, it should not overlay pads. Sometimes you want to draw information that crosses pads purely for viewing. You can do this by drawing in the tDocu layer. For example, you might draw a resistor's outline in tPlace, but draw its component leads in tDocu to make the connections more obvious on the screen and in printouts. However, when generating the silk-screen layer, the leads will not appear since tDocu is not normally produced as part of the silk screen.

Additional layers of interest include the following:

- **tRestrict, bRestrict, vRestrict** Draw shapes (polygons, circles, or rectangles) in these layers to prevent the autorouter from placing traces in the top, bottom, or via area where the shapes exist.

- **tKeepout, bKeepout** Place shapes in these layers to indicate the physical limits of the component (on the top and bottom layers, respectively). For example, a connector's body may be much larger than its pads would indicate. You might place a resistor on the board some distance from the pads, but when the board is made, you'll find out the resistor conflicts with the plastic body of the connector. With a keep-out rectangle, Eagle can tell when you've placed another component too close to the connector.

- **tNames** Insert the string >NAME in this layer to indicate the position and font for the name associated with the component. This controls how the name appears on the board layout.

- **tValues** Insert the string >VALUE in this layer to indicate the position and font for the value associated with the component.

You can also change the package's description by clicking the description link near the bottom of the editing window. This description uses Eagle rich text (see the sidebar).

Eagle Rich Text

In several places, Eagle will accept rich text—text with formatting. The formatting is very similar to HTML, but some slight differences exist. If the first line of your text contains a rich text tag, Eagle assumes you meant for it to be rich text. If the first line doesn't contain a tag, you need to start the text with a qt tag (and end it with a /qt tag). You can also use the following attributes with the qt tag:

- **bgcolor** Background color (either a named color or an HTML color attribute)
- **background** A background image

- **text** The default text color
- **link** The color for links

For example, you might write:
`<qt bgcolor="black" text="white">Fancy Text!</qt>`
You can use other tags to format your text:

`<h1> ... </h1>` A top-level heading.

`<h2> ... </h2>` A sublevel heading.

`<h3> ... </h3>` A sub-sublevel heading.

`<p> ... </p>` A paragraph. You can use the align attribute to adjust alignment (the default is left alignment; other choices are right and center). For example: `<p align="center">Custom Packages</p>`.

`<center> ... </center>` A centered paragraph.

`<blockquote> ... </blockquote>` An indented paragraph.

`<ul> ... </ul>` An unordered list. You can specify a bullet type as an attribute (for example, `<ul type="disc">`. Bullet types can be disc, circle, or square, and disc is the default.

`<ol> ... </ol>` An ordered list. You can use the optional type attribute to define the list's style. The default is type="1" for numeric labels, "a" for lowercase letters, or "A" for uppercase letters. For example: `<ol type="A">`

`<li> ... </li>` A list item within an `<ol>` or `<ul>` tag.

`<pre> ... </pre>` For blocks of code. White space is respected within the block.

`<a> ... </a>` A link or bookmark. You can use the following attributes with a link:

- **name** The bookmark name, as in `<a name="123"> ... </a>`. This defines a place in the document that other links can specify.
- **href** The link's target document. You can also specify an optional bookmark within the specified target document by placing the bookmark name after the # symbol. For example: `<a href="target.qml#123"> ... </a>`.

`<i> ... </i>` or `<em> ... </em>` Italic font style.

`<b> ... </b>` or `<strong> ... </strong>` Bold font style.

\<u\> . . . \</u\> Underlined font style.

\<big\> . . . \</big\> A large font.

\<small\> . . . \</small\> A small font size.

\<tt\> . . . \</tt\> or \<code\> . . . \</code\> Indicates monospaced text.

\<font\> . . . \</font\> Customizes the font size, family, and text color. You can use these attributes:

- **color** The text color, such as color="red" or color="#FF0000."

- **size** The logical size of the font. Logical sizes 1 to 7 are supported. The value may either be absolute, such as size=3, or relative to the current size (for example, size=-2 or size=+1).

- **face** The family of the font, such as face=times.

\<img\> An image. This tag understands the following attributes:

- **src** The image name, such as \<img src="qt.xpm"\>. Eagle supports bmp, pbm, pgm, png, xbm, and xpm files.

- **width** The width of the image. Eagle will scale the image to fit if necessary.

- **height** The height of the image. Eagle will scale the image to fit if necessary.

- **align** Determines image placement. By default, Eagle treats an image like a normal character. Specify align=left or align=right to place the image at one particular side.

\<hr\> A horizonal line.

\<br\> A line break.

\<nobr\> . . . \</nobr\> No break. Prevents word wrap.

\<table\> . . . \</table\> A table definition. This tag allows the following attributes:

- **border** Set to 0 for no border (the default) or to a number if you want to set the width of the border.

- **bgcolor** The background color.

- **width** The table width. This is either absolute in pixels or relative in percent of the column width, such as 80 percent.

- **cellspacing** Additional space around the table cells. The default is 2.

■ **cellpadding**   Additional space around the contents of table cells. The default is 1.

<tr> . . . </tr>   A table row. Can only be used within a table tag. You can use the optional bgcolor attribute to set the background color.

<td> . . . </td>   A table data cell that can only be used within a <tr> tag. You may use the following attributes:

■ **bgcolor**   The background color.

■ **width**   The cell width. This is either absolute in pixels or relative in percent of the entire table width.

■ **colspan**   Defines how many columns this cell spans.

■ **rowspan**   Defines how many rows this cell spans.

■ **align**   Alignment (left, right, or center).

<th> . . . </th>   A table header cell. Like <td> but defaults to center alignment and a bold font.

<author> . . . </author>   Marks the author of this text.

<dl> . . . </dl>   A definition list.

<dt> . . . </dt>   A definition tag that can only be used within a <dl> tag.

<dd> . . . </dd>   Definition data that can only be used within <dl>.

In addition, rich text understands several special characters that are similar to those found in HTML (see Table 5-2).

Component Creation

Creating a component involves matching one or more symbols to a package and defining the relationships between them. In many cases one symbol is assigned to a package. For example, a resistor has one package and one symbol. Of course, one library may contain many resistors (1/4W horizontal,

Table 5-2

Special rich text characters

Tag	Meaning
&lt;	<
&gt;	>
&amp;	&
	nonbreaking space
&auml;	ä
&ouml;	ö
&uuml;	ü
&Auml;	Ä
&Ouml;	Ö
&Uuml;	Ü
&szlig;	β
&copy;	©
&deg;	°
&micro;	m
&plusmn;	±

1/4W vertical, surface mount, and others). But each individual package has only one symbol (in fact, each package probably has the same symbol).

However, some components have multiple symbols per package. For example, suppose you have a dual flip-flop (like a 7474). Each package has two flip-flops, so you have two separate symbols but only one package.

When you have the package and symbols you need, you can create a new device (using Library | Device from the menu). You can click the description link (lower-left corner) and change the description of the device. Then you click the New button and select a package. You can click on the Add toolbar button to select a symbol (or more than one symbol, if applicable).

You'll use the Connect button to connect the pins on the symbol to the pads on the package. By default, the symbols will be named G$1 (and G$2, G$3, and so on) although you can change this using the Name command.

You'll also want to set the swap level and add level for each symbol (you do this by using the Change command). You've already seen how the swap level works when drawing schematics. Symbols with a swap level of zero are unique and can't be swapped. Symbols with the same swap level number are interchangeable. So if you are drawing a dual flip-flop *integrated circuit* (IC), you might make both flip-flop symbols use swap level 1. This would allow the Gate Swap command to interchange the two symbols freely.

The add level is something you haven't directly observed yet, although you've seen it in action while building a schematic. The add level has several choices:

- **Next** When you place a device, Eagle will show one of the symbols marked with Next. Each time the user runs the Invoke or Add commands, Eagle will get another symbol marked Next and place it (until no more are left). When the user adds a device and no more gates exist, Eagle places a new device. This is the default behavior and is appropriate for most devices, including those that have only one symbol.

- **Can** Gates marked with Can are not automatically placed by Eagle. The user has to run the Invoke command to select and place them.

- **Must** Some devices have multiple symbols that represent different internal items. For example, a relay symbol may have a coil and several contacts. If the relay appears on the schematic at all, the coil is required (and therefore has an add level of Must). The contacts, on the other hand, would be marked Can. So the first time you add the relay, Eagle would place the coil and one contact. You could use Invoke to add new contacts until no more could be added.

- **Request** This is similar to Can, but is only for power supply symbols.

- **Always** Gates marked always are placed on the schematic as soon as the user places the component. He or she can always use Delete to remove any unwanted symbols.

You may wonder what the difference is between Can and Request. Suppose you have an op amp with four separate amplifiers in one package. You'd probably mark each amplifier with Next, but suppose you marked them Can instead. Then the person drawing the schematic can run Invoke to pick individual gates. If the device has a name of IC1, the gates will get names like IC1A, IC1B, and so on.

Remember that Eagle connects supply nets automatically. If the IC has a supply pin named V+, Eagle will automatically connect it to the V+ supply. However, the schematic may not have a V+ supply. In that case, the user might invoke the supply symbols so they can be connected to a different supply. You don't want the supply symbols marked IC1C! That's the difference between Can and Request—Request doesn't mark the symbol as a gate in a multiple-gate package.

Loose Ends

We haven't discussed three other items on the device editor. The Prefix button, as you might guess, lets you set the prefix used for the device. For example, ICs are typically labeled IC1 and IC2, and resistors are R1 and R2. Therefore, an IC component would have a prefix of IC and a resistor has a prefix of R.

Sometimes it is necessary to have two pins with the same name (for example, an IC might have more than one GND pin). In this case, you can place an at sign after the pin name and a sequence number. For example, you might have pins named GND@1 and GND@2. These will both appear a GND in the schematic, but Eagle knows they are not the same pin.

Another item you may have noticed is next to the device description. It is marked Technologies. This allows you to have similar devices that are in different families. For example, you might have a 74L00, a 74LS00, and a 74ALS00. You can handle this by naming the device 74*00. You can select a package and then click Technologies to define the technology type (L, LS, or ALS, for example). Then the user can pick a technology when placing the part and Eagle will use the correct package.

The final item on the editor is the Value on/off radio button. If you set this radio button to on, the user can change the value of the component. A resistor, for example, usually has a value that is not known until the user draws the schematic. If you set the button to off, the user can't easily change the value (he or she can still override the value if desired, but Eagle will warn that the component's value isn't supposed to be changed). In many cases, the value of an IC is its part number (74ALS00, for example). There is no good reason for the user to change this in most cases.

A Simple Example

Consider a three-terminal resonator (a component similar to a crystal with built-in capacitors) made by Murata. The CSTLS series resonator data sheet (see http://search.murata.co.jp/image/A07X/TSMX0002.GIF) shows the physical shape of the component. The component has three pins (spaced at 2.5 millimeters or about .1 inch). The leads are, at most, 0.53 millimeters in diameter. The part's footprint is roughly 5.5 millimeters by 3.5 millimeters.

To start, create a new library. From the Control Panel, select File | New | Library.

The first step is to draw the symbol you want to use in the schematic. I decided to make the symbol look similar to a crystal, but with an extra lead (see Figure 5-4).

Select Library | Symbol from the menu and create a new symbol called RES3. Eagle will ask if you really want to create the symbol, and you should answer yes.

It is a good idea to check the grid settings before drawing anything. Usually, a .1 inch grid is a good idea and that's what this component will use.

Using the rectangle and wire commands, you can draw the basic shape of the component (see Figure 5-5) in the Symbols layer. Then you can place pins at the ends of each of the connection wires. While placing pins you can right-click to rotate the pin just as you'd rotate any component.

You probably don't want the default pin names (P$1, P$2, and P$3), so use the Name command to provide better names (1, 2, and GND). I made

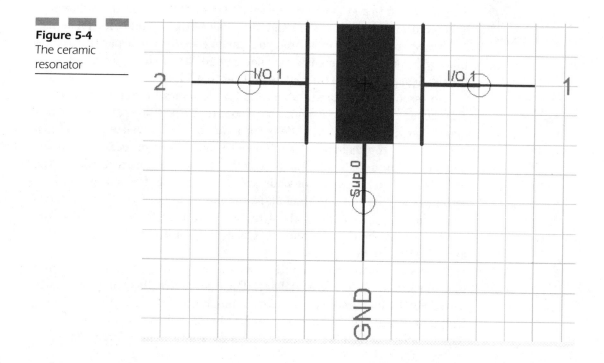

Figure 5-4
The ceramic
resonator

Figure 5-5
The basic shape
for the resonator

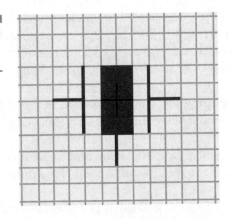

the ground pin number 3. For that matter, you don't want the pins showing on the schematic, so use the Change command to set the pin's visibility to None. The two nonground pins are interchangeable, so you should use the Change command to set the swap level of pins 1 and 2 to the same number (setting swap level to 1 will do the trick). You can also set the direction of the ground pin to Sup (for supply).

The final step is to place the parts name and value labels. Just use the Text command to add >NAME and >VALUE on the Names and Values layers to indicate where you want the component's name and value labels.

That wraps up the symbol, and you can now switch to the package (using Library | Package). Name the new package SIP3 (for single inline pin with three holes). Make sure the grid is still set to 0.1 inch.

The resonator's leads are about 0.53 millimeters in diameter or about .021 inches. A drill of .032 should be adequate. To facilitate routing, you can make elongated pads to provide extra room for traces. Select the Pad tool and from the top toolbar select YLongOct and a drill of .032. Setting the diameter to auto will allow Eagle to select an appropriate size pad for the drill size selected. Place three pads with one in the center and two more .1 inch pads to the right and the left. Name the center pad GND and the two edge pads 1 and 2 to match the symbol. You'll want to put the name and value labels in the tNames and tValues layer (just insert the text >NAME and >LABEL as before).

You also want to draw an outline wire around the part to act as an outline on the silk-screen layer (tPlace). For this, switch the grid to 0.05 inch instead of 0.1 inch and use the Wire command to draw a rectangle 6 units long and 4 units high. This is .3 inches by .2 inches (7.6 millimeters by 5 millimeters), which is a bit larger than the component, allowing for some clearance around the component. You'll also want to draw a rectangle the same size in the tKeepOut layer. Even when you are defining a surface-mount component that could mount on the bottom, always draw the component on the top layers. When the user uses Mirror to put the component on the bottom of the board, Eagle automatically flips the top layers around to the bottom. Be sure to reset the grid when you are done so you don't accidentally use the finer grid later.

The next step is to create the device (using Library | Device). Name the device CSTLS and allow Eagle to create it. Use the Add command to place the symbol in the left-hand pane. Then click New and select the SIP3 package. Leave the variant name blank and click OK. Click Connect and connect each pin on the package to the corresponding pin on the symbol.

When you return to the main editing screen, you can click Prefix (set it to X so that components of this type will be labeled X1, X2, and so on). You'll also want to click the On radio button to allow users to change the value when they place the component. Finally, you can click Description and enter an appropriate description comment.

Once you save the library, you can use it in your schematics just as you would any other library. You can use the Library | Use menu command to add a particular library to your design. You can also use the library editor to remove items from a library or rename them. After you make changes to a library you are already using, you can use the Library | Update (or Library | Update All) command to force Eagle to reload your changed library.

A Larger Example

As an example of creating a custom device, I'll show you the steps required to create a custom IC library. The IC in question is a math coprocessor (the PAK-I). It has 18 pins (see Table 5-3).

The IC is a standard 18-pin DIP, so rather than reinvent the wheel, I started by cutting and pasting the DIL18 package and symbol from the standard libraries. Be sure to enable all the layers before copying.

The next step is to set the name of each pin to match the list in Table 5-3. I also made pin 3 a clock pin (this places a V-shaped mark behind the pin). Finally, I used the Text tool to add labels for each pin and the part designator (or the Values layer).

That's the hard part. Once the symbol and the package are in place, it is a simple matter to create the device and connect the pins from the symbol

Table 5-3

PAK-I pins

Pin	Name	Description
1	SIN	Serial input
2	SOUT	Serial output
3	CLK	Serial clock
4	RESET	Reset (active low)
5	Vss	Ground
6	B0	Port 0
7	B1	Port 1
8	B2	Port 2
9	B3	Port 3
10	B4	Port 4
11	B5	Port 5
12	B6	Port 6
13	B7	Port 7
14	Vdd	+5V Power
15	RES1	Resonator
16	RES2	Resonator
17	Enable	Enable input/busy output
18	Busy	Busy output/mode select

to the package pads. In this case, you want the prefix as IC and the value radio button set to off. You can enter descriptions as appropriate. The final result appears in Figure 5-6.

In Summary

Libraries are the key to a successful PCB CAD program. I've used other programs that had limited libraries or didn't allow you to create your own

Figure 5-6
The PAK-I
component
complete

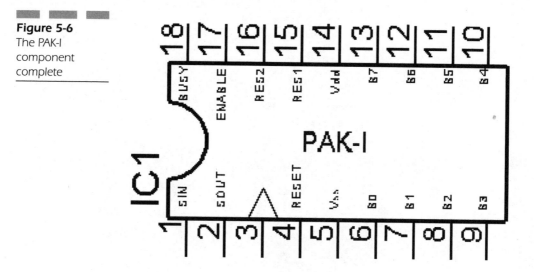

libraries. I've quickly discarded these programs. For serious work, you want a large library. But no matter how large the library is, you'll always have some part you have to work with that isn't in the library.

Eagle makes it relatively easy to define your own libraries. In fact, I often create a custom library that has the parts I use most often with any customizations that I prefer. This speeds up layout considerably because my most common parts are all in one place.

It is worth noting that Eagle only needs access to the library files when you add a new part to a schematic or board. The component is essentially copied into the data file, so you don't have to worry about libraries when you send files to another computer, for example.

If you can create schematics, boards, and custom libraries, you are set to create even the most complex board. The only piece left is to generate output ready for production. However, before you produce output, you need to learn a little about scripting, which is the topic of the next chapter.

Scripting and Programming

When I first developed an interest in computers, virtually no computers could be found in people's homes. The first computer I ever actually personally owned was an old 1802 computer with 256 bytes (that's bytes, not kilobytes) of memory. We have calculators in the house today that have far more memory and processing power. The computer at my desk is probably more powerful than all the computers I worked on prior to 1990 put together. It certainly has more memory.

There is no doubt that there have been many advances in computers over the last two dozen years. However, I'm always shocked that in one area the industry seems to have stalled or even moved backwards. That area is in scripting.

Unix users have long had the ability to control their environment via scripting. Even with graphical programs, methods (such as Tk/Tcl) allow sophisticated users to automate repetitive tasks. DOS batch files are a mere echo of Unix shell scripts—they lack the power and expressiveness. Windows users do have the Windows Scripting Host, but it is virtually unknown, and often disabled because of the threat of virus programs using it to wreak havoc.

Since Eagle can run under Linux (a system that is extremely Unix-like) it isn't surprising that it has similar scripting capabilities. You can create scripts that can duplicate any functions you can perform with the mouse. These scripts can be useful for performing precise tasks. In addition, Eagle occasionally generates scripts for your use.

Scripts are more or less a collection of commands—almost a recording of what you would do with the mouse and keyboard to perform some task. This is powerful, but sometimes you need more. For those cases, Eagle also provides a user language that is reminiscent of C. This programming language is much more powerful than scripting and can do varied tasks such as file format conversion or modifying all elements in a schematic or board.

The Command Line

Intrinsic to the idea of scripts is the command line. Up to now, you haven't used the command line for Eagle much, but it has always been there. Just under the toolbar is the command line. You can do everything you've done in previous chapters using the command line, if you want to.

Every command has a keyword. For example, the change command is, as you'd expect, CHANGE and the name command is NAME (case isn't important, so you could use change and name just as well). If you wanted to change the name of a component, you could issue the NAME command like this:

```
NAME IC4 ( 0.5 1.1) ;
```

This command changes the component at coordinate 0.5, 1.1 to use the name IC4. The semicolon is optional, but it is required for some commands so you might as well get used to using it. If you want to refer to the current cursor position, you can use @ to refer to the current cursor position. To change the component at the current position, you'd write the following:

```
NAME IC4 (@)
```

You can even mix mouse and text commands. For example, you might enter NAME IC4 and then click the component you want to change. Conversely, you could click the NAME command in the toolbar and then complete the command using the command line. If Eagle needs more information, it will prompt you. For example, if you enter the NAME command and then click a component, a dialog box will appear asking you for the name you wish to use.

The command line is useful when you want to precisely place coordinates without having to finesse the mouse. For example, suppose you want to create a new board outline that is exactly 3 by 4 inches. You could issue these commands:

```
GRID in .1 ;
LAYER dimension ;
WIRE 0 (0 0) (3 0) (3 4) (0 4) (0 0) ;
```

The online help provides details of all the commands. You can also type HELP followed by the command name to bring up the help. Table 6-1 shows the commands available. Of course, some of these commands are only valid during certain operations. For example, you can't start the autorouter while editing a schematic.

Another advantage to the command line is its history. You can use the arrow keys to scroll backwards and forwards through a list of commands you've already entered. You can edit the commands, if necessary, and reissue them.

Table 6-1

Eagle
commands

Command	Description
ADD	Adds component
ARC	Draws arc
ASSIGN	Assigns key function
AUTO	Starts auto router
BOARD	Creates board from schematic
BUS	Draws a bus
CHANGE	Changes a parameter
CIRCLE	Draws a circle
CLASS	Sets or defines a net class
CLOSE	Closes an editor window
CONNECT	Assigns a package pad to symbol pins
COPY	Copies objects
CUT	Places objects in the paste buffer
DELETE	Deletes objects
DESCRIPTION	Defines the description of a library, package, or device
DISPLAY	Shows or hide layers
DRC	Runs the design rule check
EDIT	Loads a file for editing
ERC	Runs the electrical rule check
ERRORS	Shows or hide the DRC errors
EXPORT	Exports data to another file format
GATESWAP	Swaps equivalent gates
GRID	Defines the grid units and spacing
GROUP	Defines a group
HELP	Requests help
HOLE	Adds a drill hole
INFO	Views information
INVOKE	Gets a specific device from a library
JUNCTION	Places a junction on a net
LABEL	Labels a bus or net
LAYER	Changes and defines layers
MARK	Sets the relative mark
MENU	Customizes the command menu
MIRROR	Mirrors a component
MOVE	Moves an object
NAME	Sets the name of an object
NET	Defines a net
OPEN	Opens a library for editing

Command	Description
OPTIMIZE	Joins wire segments
PACKAGE	Defines a package variant
PAD	Adds a pad
PASTE	Inserts the contents of the paste buffer
PIN	Defines a pin on a symbol
PINSWAP	Swaps equivalent pins
POLYGON	Draws a polygon
PREFIX	Defines a symbol's prefix
PRINT	Prints the document
QUIT	Exits the program
RATSNEST	Recalculates airwires
RECT	Draws a rectangle
REDO	Repeats an operation previously canceled with Undo
REMOVE	Removes an item (files, library items, and so on)
RENAME	Renames symbols, devices, or packages
REPLACE	Replaces a package on a board
RIPUP	Rips up a trace
ROTATE	Rotates a component
ROUTE	Defines a route
RUN	Executes a user language program
SCRIPT	Runs a script
SET	Sets system options
SHOW	Shows information
SIGNAL	Defines signal
SMASH	Separates label from component
SMD	Adds an SMD pad
SPLIT	Splits wires and polygons into segments
TECHNOLOGY	Defines technology names
TEXT	Enters text
UNDO	Undoes an operation
UPDATE	Updates library objects
USE	Specifies a library to use
VALUE	Sets a component's value
VIA	Adds a via
WINDOW	Zooms in or out
WIRE	Draws a wire
WRITE	Saves the current document

Using User Language Programs and Scripts

Eagle ships with a variety of scripts and *user language programs* (ULPs) that perform useful functions. In addition, you can export certain items as scripts that you can execute to recreate the item.

For example, suppose you are finished with a board and you call some board makers to ask them how much it will cost to produce the board. They'll want to know the size of the board and the number of layers—those are easy to figure out. However, another data point they'll want is the number of holes to drill. On a complex board, this will take some time to manually count. Luckily, a ULP (specifically COUNT.ULP) does the job for you.

To run the ULP, simply select File | Run from the main menu and pick COUNT.ULP from the resulting file dialog. Of course, you must do this while the board in question is open. Figure 6-1 shows a typical result. Once you dismiss the dialog box, the user program will offer to save the results to a file. If you don't want to save it, simply click Cancel.

Many useful ULPs are available in the Eagle ULP directory. For example, the BOM.ULP program will prepare a bill of materials from a schematic and the SILK.ULP program will widen all silk screen on a board (necessary when your board manufacturer can't produce narrower silk-screen markings). You can browse the available programs (which include verbose descriptions) using the Eagle Control Panel. ULPs are just text files, and you may have to modify certain parameters at the start of the file to customize the program's operation. Programs that require customization

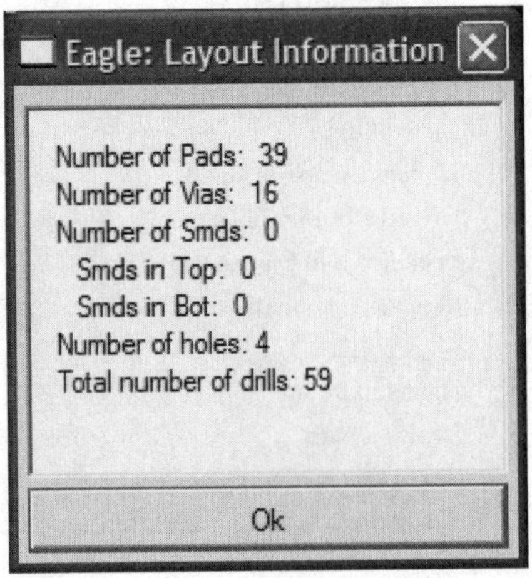

Figure 6-1
The COUNT.ULP program counts the number of holes in a board.

Eagle: Layout Information

Number of Pads: 39
Number of Vias: 16
Number of Smds: 0
   Smds in Top: 0
   Smds in Bot: 0
Number of holes: 4
Total number of drills: 59

Ok

should include complete instructions. For example, here are a few lines from the start of the SILK.ULP file:

```
                                    // Define your own silk-screen width
here
real Silkwidth   = 8.0 ;    // in mil
int  NewTextRatio = 16 ;    // and the new text ratio

int  source, newlay, tplace = 21,
                      bplace = 22,
                      offset = 100;
```

The first two items are what you usually want to change, the silk-screen width and the ratio for text. Unfortunately, the next items are not commented on by Eagle. This line tells the program to consider layers 21 and 22 as the silk-screen layers. The offset parameter causes the program to output the new layers in layer 121 and 122. You might change the offset to 0 if you want to write the new silk screen over the original layer.

CADSoft's web site also has a variety of user-contributed files available. For example, the Restrict.ULP file will draw restrict pads around all pads (useful when you plan to use jumpers for the top layer of the board as discussed in Chapter 5, "Custom Libraries").

Script Files

Creating a script file is simple. A script file is nothing but a text file full of commands. If you select File | New | Script from the Control Panel, Eagle just opens an empty text file. You can use Notepad or any other plain text editor you prefer. By default, scripts have a .SCR extension. You can put comments in the script by starting them with the # character. Eagle ignores everything from that character to the end of the line.

The file contains the usual things you'd type in from the command line. Suppose you want to create an outline for a board that takes up half a Euro card (a particular panel size often used by board makers and, incidentally, the maximum size board you can create with Eagle's free edition). You can create a script like the one in Listing 6-1.

Listing 6-1

A script to create a board outline

```
# Half Euro Card
Grid mm 1 off;
Set Wire_Bend 0;
Layer Dimension;
Wire 0  (0 0) (100 80) (0 0);
```

(continued)

Listing 6-1
(continued)

A script to create
a board outline

```
Layer Top;
wire 2  (20 -1) (-1 20);
wire  (80 -1 ) (101 20);
wire (20 81) (-1 60);
Set Wire_Bend 4;
wire (101 60) (80 81);
Layer Bottom;
Set Wire_Bend 0;
wire 2  (20 -1) (-1 20);
wire  (80 -1 ) (101 20);
wire (20 81) (-1 60);
Set Wire_Bend 4;
wire (101 60) (80 81);
Grid Last;
Window Fit;
```

Customizations

In addition to automating common tasks, you can also use scripts to customize Eagle. When Eagle opens an editor window (or creates a new one), it reads the EAGLE.SCR file (in the current directory or the system-defined script directory). This file allows you to execute scripts for each type of editor (except the text editor). For example, consider this excerpt from an EAGLE.SRC file:

```
BRD:
Menu Add Change ;
```

The BRD: label indicates that this section of the file applies to the board editor. You can use SCH: (schematic) or LBR: (library). In addition, you can provide DEV: (device), SYM: (symbol), and PAC (package) labels to influence those editors. The Menu command controls the menu bar at the right side of the screen. You probably have this bar hidden. To show it, select Options | User Interface from the main menu and check the Command Texts box. Keep in mind that you can click on the double lines near each toolbar or menu to hide it or show it, and you can also move these items around on the screen. So although the default position for the buttons is on the right-hand side of the screen, you may want to move it to another area.

The previous example changes the default menu buttons to have only two items: Add and Change. You can also make a custom label and associate it with one or more commands:

```
Menu 'NormalView : Display None Top Bottom Pads Vias;'
```

This creates a button with the name of NormalView. The button executes the commands listed. Here is another example:

```
Menu 'RouteIt: Grid mm 1 ; route '
```

This creates a button named RouteIt. Pushing this button sets the grid to 1 millimeter and also begins the Route command. Notice that no semicolon is used at the end of the Route command. That's because you want to complete the Route command with the mouse. If you enter the semicolon in the menu definition, it will terminate the Route command, which makes no sense.

Buttons can even have subbuttons. Consider this example:

```
Menu 'SetGrid { 1mm : Grid mm 1; | 5mm : Grid mm 5; }';
```

This created three buttons. The main button contains the label SetGrid. When you push it, two more buttons, 1mm and 5mm, appear.

Of course, the EAGLE.SCR file can contain any commands, not just menu settings. For example, you might want to set a default grid or use a custom library upon startup.

Creating User Language Programs

Sometimes scripts just aren't enough to perform sophisticated tasks. That's where ULPs are handy. ULPs are interpreted programs that use a programming language akin to C. The ULP can access and modify nearly all aspects of Eagle documents. This makes them a powerful tool for converting to other file formats, collecting information, or making mass changes to a schematic or board.

A complete treatment of ULPs would require a complete book in itself. However, the rest of this chapter will give you the fundamentals required to create your own ULPs. You can find complete details about ULPs in Eagle's online help files. If you want to write a ULP, it will be helpful to have experience with object-oriented programming languages. If you don't feel like you'll need to write ULPs, you can safely skip the rest of this chapter.

A Basic ULP

Listing 6.2 shows an extremely simple ULP. The #usage directive documents the program and the associated text appears in the Eagle Control Panel.

The next line defines a string variable (ttext) that contains an arbitrary string. The final line displays this string variable in a message box.

Listing 6-2

*A very simple
ULP*

```
#usage "Just a test program"
string ttext = "I'm Eagle!";
dlgMessageBox(ttext);
```

ULPs support several types of variables:

- **char**   Single characters
- **int**   A 32-bit integer
- **real**   A floating-point number
- **string**   A series of characters

If you save the text in Listing 6-2 in a file named test.ulp, you can execute the program from the command line by entering

```
RUN test.ulp
```

Alternately, you can use the File | Run command from the menu.

Most ULPs will also use functions. For example, Listing 6-3 shows a more complex example. This program defines a function named double. This function accepts a real number, multiplies it by two, and returns a real value. It is important to realize that the function definition doesn't cause any code to execute until a call is made to the function. If no part of the program calls the double function (as the two sprintf statements do in the main part of the program), then the function will never actually do anything.

The main part of the program uses sprintf to format a string variable. Consider this line:

```
sprintf(ttext,"Double 10 = %f",double(10));
```

This statement calls the double function with an argument of 10. This, of course, returns 20. The sprintf function replaces the %f in its second argument with the 20. The resulting string winds up in the ttext variable. The next line displays the resulting string in a message box just as Listing 6-2 does. Then the program repeats the steps with a different number. This is a contrived example, of course, but it illustrates how you can combine repeated operations into a function.

Listing 6-3

A more complex example

```
#usage "Just a test program"

string ttext;
real double(real r)
{
   return r*2;
}

sprintf(ttext,"Double 10 = %f",double(10));
dlgMessageBox(ttext);

sprintf(ttext,"Double 3.39 = %f",double(3.39));
dlgMessageBox(ttext);
```

Operating on Documents

Each Eagle document consists of an object that contain data and the operations that you can perform on the data. Each object contains other objects that represent the items within the document (and these objects can, in turn, contain even more objects). For example, suppose you want to see information about the grid size of the current board. You might write the following:

```
board(b)
  {
  sprintf(ttext,"%f",b.grid.distance);
  dlgMessageBox(ttext);
  }
```

The first line creates an object variable, b, that refers to the current board. This object is of type UL_BOARD. The notation b.grid refers to a grid object that is within the board. The grid object has a member, distance, that tells you the size of the grid.

You can look up each object type in the online help. The top-level objects are of type UL_BOARD, UL_SCHEMATIC, and UL_LIBRARY. Just as an example, a board contains a UL_AREA, a UL_GRID, and a name (which is simply a string). It also contains a collection of objects that refer to the individual elements (for example, UL_LAYER, UL_HOLE, and UL_SIGNAL objects). In addition to the three main root object types, you can also retrieve objects that represent other items such as sheets.

In a collection, you have to loop through the list of objects to find the one you want. For example, suppose you want to display the name of each signal on the board. You might write

```
board(b)
  {
    b.signals(s)
```

```
    {
    dlgMessageBox(s.name);
    }
}
```

Often you'll want to know if the current editor is working with the type of document you expect. For example, if you wanted to make sure the editor had a board open, you could write

```
if (board)
  {
  board(b)
    {
    . . .
```

The if statement only executes the following group of statements if the condition in parenthesis is true. In this case, the condition is only true if a board is loaded. However, sometimes you want to refer to, for example, the board associated with an open schematic. You can accomplish this by using the project prefix. Here's an example:

```
project.board(b)
  {
  . . .
  }
```

Command and Control

The if statement is just one of the C-like control statements you can use in a ULP. Any block of statements surrounded by curly braces is treated as a single statement. Table 6-2 shows the available control statements.

You can use many built-in subroutines, statements, and functions. You'll find them all in the online help. Certain functions can manipulate strings, work with files, print, and handle other common chores.

If you dig through Eagle long enough, you'll find that internally Eagle represents measurements in .1 micron units. Functions can convert this internal format to more common units (like mils and millimeters) and back.

Far more functions and statements exist than will fit in a single chapter. If you know a bit of programming, you'll be able to get the rest from the online help. Also, one of the best ways to learn about ULPs is to read through some of the examples provided in the Eagle ULP directory.

Table 6-2

Control
statements

Statement	Description
do . . . while	Executes a loop at least once and continues while a condition is true.
for	Executes a loop, typically for a number of times.
while	Executes a loop while a condition is true (will not execute if condition is initially false).
if . . . else	Conditionally executes statement.
switch	Selects one of multiple conditions.
break	Exits a loop early.
continue	Returns to the top of a loop.
return	Returns from a subroutine.

In Summary

Scripting in Eagle is a powerful way to automate repetitive tasks. In addition, scripting allows you to customize Eagle's menus and provide custom commands when you open different types of documents.

As you gain experience with the scripting interface, you'll find you appreciate the precision the command line interface affords you. Even without scripting, you'll find it convenient to use commands to precisely specify your desires to Eagle.

If you are a programmer, you'll appreciate Eagle's ULPs. Even if you aren't comfortable programming, you'll find many of the predefined ULPs are useful or even essential. For example, you'll use a ULP to help prepare output files to send to a board manufacturer, the subject of the next chapter.

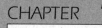

Eagle Output

I recently showed a friend of mine one of my books, and he commented that he liked the layout. It surprised him to find out that I have practically nothing to do with their layout. In this day of the modern word processor, you'd expect authors to write books with sophisticated styles and formatting. The publisher could then just produce the book as written. You might expect that, but it hasn't been true of any of the books (or magazine articles) that I've written.

The publishers hire experts who will do a much better job of laying out a book than I ever could. They use high-end software packages to perform the page layout. However, that means they want practically no formatting in the original manuscript. The first few books I ever wrote required me to submit printed pages, but at least those days are gone. Now, however, you do have to turn in plain vanilla files with almost no frills.

If I printed my manuscript on a laser printer, it wouldn't look nearly as good as what you have in your hands right now. It takes a lot of effort to finesse professional-looking output. When you have a PCB designed in Eagle, it is much like a manuscript in a word processor. The idea is there, but you still have to produce some sort of output.

In some cases, you'll want to create output so you can actually build your board yourself. In other cases, you'll want to produce a set of files that a board maker can use to professionally produce your board.

Eagle can handle either case. The steps are almost identical, except for the target of your efforts. If you are making your own boards, you probably want to print to a printer or produce a plot using a plotter. If someone else is making your boards, they probably expect Gerber files, but some board makers will accept Eagle BRD files directly. CadSoft allows them to use the free version of Eagle to produce the output for you. If you are using someone who will do this, you can essentially skip this chapter unless you simply want a better understanding of the process.

Output Methods

The final goal of an Eagle project is a finished *printed circuit board* (PCB). If you are using one of the techniques described in Chapter 8, "Boards from a Laser Printer," and Chapter 9, "Photographic Boards," you'll need a likeness of the board on a piece of paper or transparent film. These days, that usually means output to a laser printer. For some purposes, you might be able to print on an inkjet printer, as well. Another output option is to use a plotter (a computer peripheral that draws using pens), although these are much less common today than printers.

A few other less common output options are available. Some machines can mill copper away from a board under computer control (these machines are not inexpensive, as you might imagine). A professional board maker will use a machine called a photoplotter. A traditional photoplotter uses a beam of light to draw on a piece of photographic film. These traditional photoplotters are known as vector photoplotters. Modern photoploters are usually raster photoplotters that work more like a traditional laser printer, drawing dots on the film instead of a laser printer drum.

The most common brand of photoplotter is made by Gerber, and the file formats a Gerber accepts are the *de facto* standard for photoplotters. As a result, most board makers (and also some specialized devices such as PC milling machines) expect you to provide Gerber files. You may encounter other file formats, such as PostScript or *Hewlett Packard Graphics Language* (HPGL, a coding system used mainly for plotters). Eagle can prepare all of these and more.

Several different versions of Gerber files exist. Most modern equipment will accept RS274X Gerber files, and that is what most of this chapter will cover (using the GERBER_RS274X driver). However, Eagle can also create RS274D files (an older standard) through the GERBER and GERBERAUTO drivers.

Each layer of your physical board will correspond to one Gerber file. Keep in mind, however, that one physical layer may correspond to multiple Eagle layers. For example, the top layer of the board contains Eagle Top, Pads, and Vias layers. In addition to the layers, you'll probably want to create a file that describes the drills to be made in the board. A common drill machine is an Excellon, and the file format—the Excellon file format—is the most often used format for describing drills. Other formats are available (such as SM1000 and SM3000), but Excellon is by far the most common format required.

Printing

One simplistic way to print your board design is by hiding all the layers you aren't interested in and then using the File | Print command (see Figure 7-1). The Print command allows you to select several options. If you are printing a board layout, you'll probably want to select black and solid.

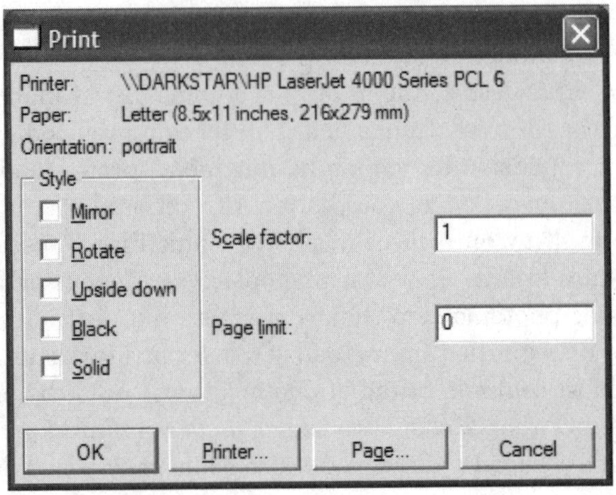

Figure 7-1
The Print
command

Unless you are doing some photographic trickery, you'll want to set the scaling to 1.

The Print command also allows you to print a mirror image of the board. This is useful in several situations. One common way of making boards is to print the pattern on paper and then transfer the image on the board like an iron-on decal. In this case, you'll want to print a mirror image of the artwork so that when you transfer the image to the blank board, it will be in the correct orientation. In addition, when printing to transparencies, you may want to mirror one side of the board so when the films are aligned, the images will be correctly oriented.

Computer-Aided Manufacturing

The Print command is adequate, but it is cumbersome to use and has limited output options. Usually, you'll want to use Eagle's *Computer-Aided Manufacturing* (CAM) processor instead. The processor allows you to create job files that can produce a group of output files at one time. You can save commonly used jobs, and Eagle provides several useful predefined jobs for you.

The CAM processor has the familiar tabbed-style interface (see Figure 7-2). Each tab corresponds to a single output file. If you open the CAM processor from the Control Panel or from the File menu, it won't appear to have any tabs. You can use the Add button in the lower left-hand corner to add more tabs (and therefore more output files).

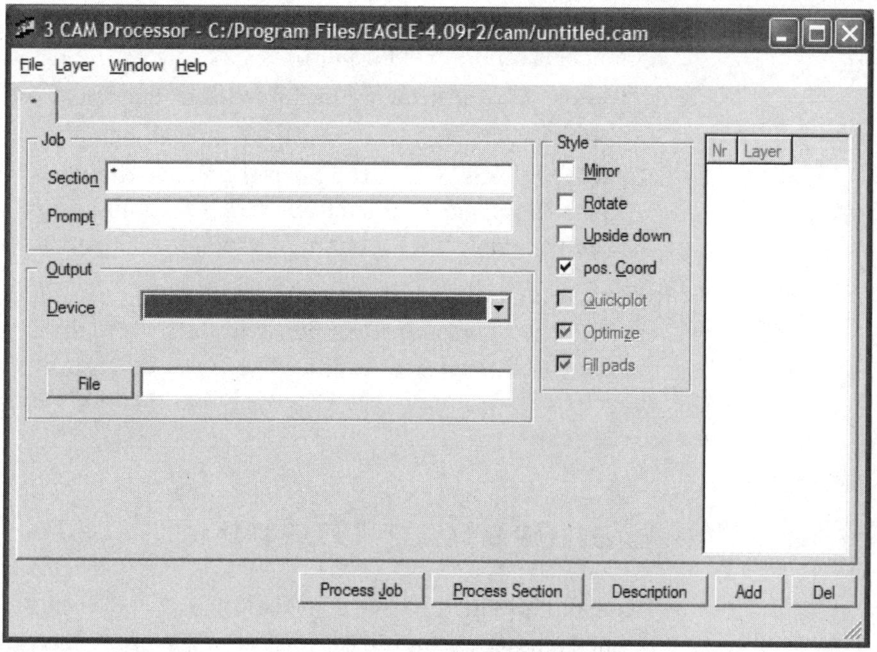

Figure 7-2
The CAM
Processor

Each tab has a number of entries:

- **Section**   The name of the tab (for example, Component side).

- **Prompt**   If this field is not blank, Eagle will prompt you with this string before processing this tab. For example, the prompt might ask you to insert a new sheet in a plotter before continuing.

- **Device**   The device driver to use. Many of these correspond to plotters or printers. Some are virtual devices like the ones that create Gerber or PostScript files.

- **File**   The output file. If the file name begins with a period, Eagle will use the basename of the file, plus the specified extension. If the File field is .cmp, and you are processing a board called demo.brd, the output for this tab will be demo.cmp. The file name can also be a device name, if required.

- **Offset**   This field allows you to shift the top left of the output by the indicated amount.

- **Style**   You can use this group of buttons to set various options regarding the output. For example, it is possible to rotate or mirror the output here. The Pos Coord box allows you to force all coordinates to be positive to avoid clipping. Optimize causes Eagle to generate more efficient commands for plotters.

- **Sheet**   If your drawing has more than one sheet you can select the sheets to use here.
- **Layers**   This box on the far right shows the layers available. Layers that will appear in the output will be highlighted. If no layers appear, then you don't have a board or schematic file open.

The Process Section button will generate the output corresponding to the current tab. Using Process Job will run through all the tabs. As you'd expect, the Description button allows you to enter or edit a description of the job (this appears in the Control Panel) and Delete removes tabs.

Generating Output

Instead of producing your own CAM jobs, you'll usually use one of the predefined jobs (or perhaps a copy of one of the predefined jobs with slight modifications). Most board houses will expect the following files (assuming the base name of the file is board):

- **board.cmp**   The component (top) side of the board. This consists of Eagle's Top, Pads, and Vias layers.
- **board.sol**   The solder (bottom) side of the board. This consists of Eagle's Bottom, Pads, and Vias layers.
- **board.plc**   The component-side silk screen. This usually consists of the tName and Dimension layers. It may also include the tValue layer.
- **board.pls**   The solder-side silk screen, which is usually comprised of the bName and Dimension layers and (optionally) the bValue layer.
- **board.stc**   Component-side solder mask (tStop).
- **board.sts**   Solder-side solder mask (bStop).
- **board.crc**   Component-side cream.
- **board.crs**   Solder-side cream.
- **board.drd**   Drilling data.

It is also customary to provide a readme.txt file that explains the layers and provides your contact information. Sometimes it is useful to create a file with just the outline of the board in it (a file with a .gko extension containing just the Dimension layer usually) to indicate how the board will be cut. Most of the time, you'll want to package these files together using a ZIP utility so that you send them as a single file.

Not every board will have all of these layers, of course. For example, a single-sided board won't have a component side or component-side solder mask. Boards that won't use solder paste won't use the cream layers.

Here are the steps I use to create my Gerbers before sending them to a board house:

1. Open the board in the board editor.

2. Use File | Run and select drillcfg.ulp (a predefined user language program).

3. Select the output units you desire (usually, board makers in the United States will want inch units and everyone else will want millimeters).

4. A dialog will appear (see Figure 7-3) showing the drill bits required to produce your board. You could potentially change these numbers to match a standard drill rack (a rack is a set of drill bits used in an automatic drilling machine) as long as the sizes are very close to the original sizes.

5. Press OK and save the drill rack file (use a DRL file extension and the same base name as the board).

6. From the board editor, use the File | CAM Processor command.

7. Use the File | Open | Job menu in the CAM processor.

8. Open the aawstandard.cam file (available on the CD-ROM).

9. Browse through the tabs, making sure the layers you want enabled are on. For example, you may want to turn on the tValues layer in the silk-screen layer (I prefer not to have the values on the board).

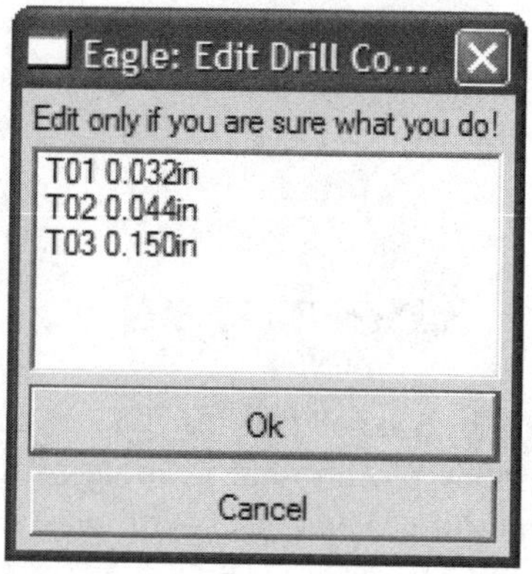

Figure 7-3
The drillcfg user language program determines the drill bits required to produce your board.

Eagle: Edit Drill Co... ✕

Edit only if you are sure what you do!

T01 0.032in
T02 0.044in
T03 0.150in

Ok

Cancel

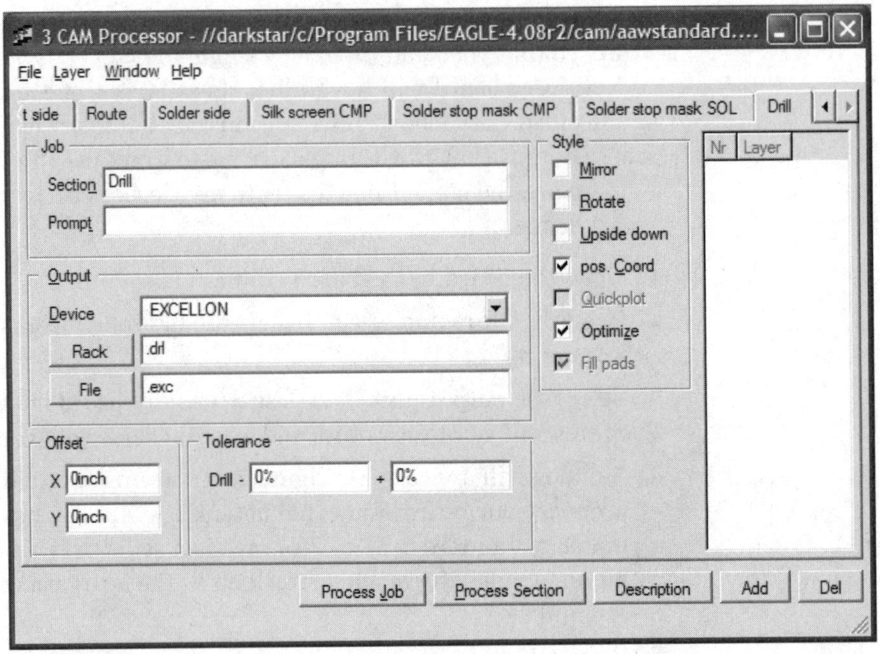

Figure 7-4
The drill output
section

Figure 7-4
The drill output section

10. Select the Drill tab (see Figure 7-4) and ensure you have the correct rack file (from step 5) selected. Remember, if the field contains just .drl, Eagle will use the base name of the board and the .drl extension.

11. Push the Process Job button.

For a two-layer board, with silk screen on one side and no solder cream layers, this is about it. You can zip up all the files created, the rack file, and a readme.txt file and simply send it to the board maker. If you have more than two layers, or other requirements, (for example, a solder-side silk screen) you'll have to add the appropriate sections (tabs) to the CAM job.

You may want to check the Gerber output to make sure it is what you expect. You'll read more about that in Chapter 10, "Outsourcing Boards."

Gerbers with Apertures

Modern photoplotters are like a laser printer. They can create any shape you define. However, older photoplotters use a wheel with pieces of film that contain various patterns on them. These patterns are called apertures. The photoplotter might have several diameters of circular apertures and several other apertures with a square appearance, octagonal shapes, and so forth. These apertures determine the shapes the photoplotter can draw on the film.

The photoplotter understands two operations: flashing and drawing. For example, suppose your board calls for an octagonal pad. The photoplotter will move to the correct part of the film, spin the wheel to an octagon aperture, and then turn the light on (and back off). This essentially makes a copy of the aperture on the film.

To make a track, however, the plotter draws the pattern. It will move to the starting point, switch the wheel to, for example, a square aperture, and then turn on the light. With the light on, the photoplotter will then move to draw the pattern of the trace. When it reaches the end, the photoplotter will extinguish the light and go to the next pattern.

Eagle will draw your board in terms of apertures. However, as you might expect, it has to know which apertures are available and their position on the aperture wheel.

If your board maker requires that you conform to a fixed wheel, you'll use the GERBER and GERBERAUTO driver (instead of the GERBER_RS274X driver). You'll also need to set the wheel file to match the setup. Your board maker can probably provide this file, or you can set it up using the information in the Eagle manual. You can ask Eagle to emulate apertures that don't exist in the wheel, although this can be crude and creates large output files. After generating this type of Gerber file, you can examine the .GPI file to see if any apertures were not available or were emulated.

Filename Tricks

Naturally, you can specify an entire filename in any of the file fields of the CAM processor. However, you'll usually just specify an extension. By default, Eagle will place the files in the same directory as the board file. If you want to specify a particular output directory, you may do so. For example, you might specify c:\camoutput\.stc (or, under Linux, /home/alw/camoutput/.stc).

One possible problem is the disposition of information files. Each time you run a section of the job, Eagle writes a .gpi file, but that overwrites the previous .gpi file. If you want to preserve the information (gpi) files, you can use a file extension with a # character in it to work around this problem.

Suppose you create a section that has a file extension of .st#. Eagle will generate the output file with an extension of .stx and the information file with .sti. If you select the file names correctly, it is possible to generate many separate information files, one for each section, without overwriting information from other sections.

Inside Gerber

Gerber files are nothing more than ordinary text files with data for the photoplotter (or other device). These files are quite simple because early photoplotters were not very sophisticated.

You can get the complete specification for Gerber files at www.maniabarco.com/transdown.asp (Gerber was bought by Mania Barco, a European company). Only four basic commands appear in a basic Gerber file. However, a basic Gerber file assumes you knew certain options were in force when it was created. One of the big advantages to the newer RS274X format is that these options are encoded into the file, which eliminates the potential for human error (the RS274X format also embeds the aperture definitions, so a wheel file is not needed).

The four types of commands (or codes) in a Gerber file are G, D, M, and coordinate data. Not surprisingly, each command starts with the letter indicated (or X or Y for coordinate data).

The G codes set the photoplotter's initial state. These are sometimes omitted, which means the operator must know the intent of the person who created the Gerber. A G90 command tells the photoplotter that the coordinates specified are absolute (the default), whereas G91 indicates that each coordinate is relative to the current position of the plotter. Likewise, a G70 command sets the measurement scale to inches and a G71 command sets the plotter to millimeter mode. All the commands end with an asterisk, so you might see

```
G90*
G71*
```

You may also see a G54 command, which instructs the plotter to move the aperture wheel to the position indicated by the next D code. Most plotters will automatically do this when they encounter a D code, so you may not find any G54 codes in your Gerber files.

The D codes actually draw (or flash an aperture). D01 tells the plotter to leave the light on (or, more likely, a shutter over the light open) and move to the indicated coordinates. D02 does the same thing, but with the light off (that is, with the shutter closed). D03 is the Flash command. It moves to the indicated coordinates, opens the shutter, and then closes it again.

These D codes follow coordinate data. So a typical section of a Gerber might look like this:

```
X0Y0D02*
X100Y50D01*
```

The first line brings the plotter to location 0,0 with the shutter closed. Then it moves to location 1.00, 0.50 with the light on (which would draw a line). Obviously, the decimal point is assumed (and where it occurs is something the operator needs to know if you aren't using the RS274X format). Also leading (or trailing) zeros can be suppressed. The plotter operator needs to know which zeros you are suppressing (if any). In addition, coordinates are only required when they are different from the previous coordinates. So suppose the file had a third line:

```
X200D01*
```

This would change X to 2.00 without changing Y and execute the specified D code.

Eagle's information file tells the operator the information about the files it generates when you use the GERBER driver. Here's an example information file:

```
Photoplotter Info File: C:/Program Files/EAGLE-
4.09r2/projects/Test/ar1-preroute.gpi
 Date : 26-03-2003 14:21:03
 Plotfile : C:/Program Files/EAGLE-4.09r2/projects/Test/
ar1-preroute.sts
 Apertures : C:/Program Files/EAGLE-4.09r2/projects/Test/
ar1-preroute.whl
 Device : Gerber photoplotter
Parameter settings:
 Emulate Apertures : no
 Emulate Thermal : no
 Emulate Annulus : no
 Tolerance Draw + : 1.00 %
 Tolerance Draw - : 1.00 %
 Tolerance Flash + : 1.00 %
 Tolerance Flash - : 1.00 %
 Rotate : no
 Mirror : yes
 Optimize : yes
 Auto fit : yes
 OffsetX : 0inch
 OffsetY : 0inch
Plotfile Info:
 Coordinate Format : 2.4
```

```
Coordinate Units : Inch
Data Mode : Absolute
Zero Suppression : None
End Of Block : *
Apertures used:
  Code Shape Size used
  D11 round 0.1580inch 4
  D13 oval 0.1120inch x 0.0600inch 8
  D18 octagon 0.0600inch 29
  D22 octagon 0.1330inch 2
```

Notice the Plotfile Info section. It contains all the information the photo-plotter needs to know (the units, the decimal point position, and the zero suppression settings). It may seem silly today to worry about removing zeros and decimal points. However, when the input to these machines was paper tape, cutting 25 percent of your data made life much easier.

Commands from D10 to D999 select specific apertures from the wheel file.

The only other common code in a Gerber file is a M02. This indicates the end of the file. Most plotters will ignore anything after the M02 code.

Changing Gerbers

Since the Gerber file format is simple, you can do things to post-process the Gerber if you are careful. For example, suppose you want to gang four copies of the same board in a 2x2 grid. You'll often want to do this (known as step and repeat) when you are making small boards on a large standard-size panel.

Most board makers will do this for you. However, it is easy enough to introduce special commands in the Gerber file that will repeat the entire data file. Suppose you have a 3-inch by 2-inch board and you want to repeat the board 4 times across and 5 times vertically. What's more, you want a .25-inch border between the boards. You could place this line at the start of your RS274X Gerber file:

```
%SRX3Y4I3.25J2.25*%
```

This causes the plotter to step and repeat (hence, the SR) 3 times in the X direction and 4 times in the Y direction. Each step will be 3.25 inches across from the previous step and 2.25 inches down from the previous step.

Incidentally, an open source program will do this for you. Have a look at http://gbtiler.sourceforge.net/. This is a Perl program so it should be portable, although you'll need Perl (another free program) to run it.

Of course, stepping the image is only part of the problem. You also need to step the drill file. You can do this by hand, although it is tedious and error-prone. If you open an exc file produced by Eagle, you'll see it isn't very mysterious:

```
%
T01
X429Y335
X529Y335
X779Y385
X779Y685
X529Y685
X429Y685
T02
X429Y1035
. . .
X1979Y1885
T03
X179Y185
X1979Y185
X1979Y2235
X179Y2235
M30
```

You can probably guess that the T commands in the file refer to tools in the rack file. Each X and Y coordinate indicates a hole to drill. M30 ends the file. Some board houses expect the tool information to appear at the start of the Excellon file, which would require you to manually modify the file:

```
M48
T01C0.032
T02C0.044
T03C0.150
%
T01
X429Y335
X529Y335
. . .
```

You could copy and paste the entire group of coordinates and manually add the offset for each repeat. However, usually the board maker will be willing to do this for you.

In Summary

The best word processor in the world isn't of much use unless you can print or otherwise publish your output. Likewise, Eagle would not be very useful if it couldn't generate output you can use to make your PCBs (or have them made). Some low-end PCB programs can only print to your printer or save data to proprietary file formats. This would be okay if you were just making your own boards, but eventually you will want to send your files somewhere to have them professionally made. At that point, you will appreciate the versatile and powerful output options available with Eagle.

If you are making your own boards (the subject of the next two chapters), you'll probably be mostly interested in printing board artwork to a laser printer or inkjet printer. If you are having boards professionally made (the subject of Chapter 10) you'll be mostly interested in Gerber output. Either way, the CAM processor is the answer to getting exactly the files you want.

You'll read more about Gerber output—and particularly previewing Gerber output—in Chapter 10. If you don't plan to make your own boards, you can skip directly to that chapter now.

Boards from a Laser Printer

You could make the argument that the Protestant Reformation—or even the Renaissance in general—was brought about by the printing press. Prior to the printing press, there wasn't much reason for ordinary people to learn to read. Even people who did read didn't have a lot to read. A library from the Middle Ages looked more like a bank vault than a public library. Hand-copied books were so valuable, libraries throughout Europe chained their books (Hereford Cathedral in Britain still has its chains and fixtures).

The movable type printing press, which Gutenburg patterned after a wine press, allowed literate people to exchange books (and, more importantly, ideas). It also provided greater motivation for literacy because the average person had greater access to printed material. Although the Chinese used movable type printing earlier than Gutenburg, the Chinese language with its many characters still made printing difficult compared to the easily standardized European presses.

Surprisingly, the printing press remained more or less unchanged until 1814—almost 370 years after its introduction. The next 200 years saw printing change radically, however. The typewriter and mimeograph were the first to bring printing, of a sort, to the desktop. Xerography changed the office forever. And, of course, computer printers grew up from giant line printers to dot matrix, inkjet, and daisy wheel printers.

In 1978, Xerox introduced a computer printer that worked on the same principle that a photocopier uses (this giant printer could print 120 pages per minute when it hit its stride). This laser printer would become the gold standard for printing and 10 years later, desktop laser printers were reasonably affordable. Ten more years would see laser printers become commonplace and relatively inexpensive.

What does this have to do with *printed circuit boards* (PCBs)? Laser printers are an excellent way to output artwork for printed circuits. Also, it is possible to use a laser printer to create PCBs.

The basic way to make a PCB is to start with a copper-clad board. You can draw your artwork—somehow—on the board using some *resist*—a chemical that will cover and protect the copper where you expect to have traces. The rest of the board is left bare. Then you can use a chemical to remove the copper where the resist is not present. When you are done, you remove the resist and you have a board ready for drilling.

The trick is how to transfer your design to the board. Several possibilities exist:

- Use a permanent marker and draw by hand.
- Use special rub-on symbols designed for this purpose.
- Use a computer plotter to draw the resist automatically.

- Use a photographic process to print the resist pattern (the subject of Chapter 9, "Photographic Boards").

- Use a laser printer to transfer toner to the board.

Obviously, permanent markers aren't going to give you great results—especially with complex artwork and fine traces. The rub-on symbols are fine for a single board, but you can't really mass-produce boards this way. Computer plotters are good, but not very common these days. Besides, you have to find pens that have ink that will resist the etching process, or make your own pens from permanent markers.

The photographic process is closest to the way boards are made professionally. It gives you great results, but it is somewhat difficult and messy as you'll see in Chapter 9. If you already have a laser printer, you'll find it can save you a lot of time and produce excellent boards.

Even if you don't plan on using the laser printer to produce boards, you'll still want to read the section in this chapter about drilling and etching boards. These steps apply no matter which method you use to put the resist on the blank board.

Laser Fundamentals

You might wonder how you can print a resist pattern on a copper board using a laser printer. After all, I doubt a printer can feed a normal copper board (although I have heard of people printing on thin copper foil, I wouldn't recommend it). The trick relies on how laser printers work.

The basic technology behind photocopying and laser printing was invented in 1938. The inventor, Chester Carlson, tried to sell the idea to RCA, IBM, Kodak, GE, and many other companies. None of them thought it was a good idea.

A laser printer uses a laser (or sometimes a *light-emitting diode* [LED]) to draw dots (in a mirror image of the desired page) on a special drum. Where the laser touches the drum, the drum develops an electrical charge. The drum turns and passes through toner—pulverized black plastic. Just as socks cling together, the toner sticks to the drum where the surface has a charge. As the drum continues to turn, it presses against the paper and the toner sticks on the paper, forming the printed page.

If the process stopped here, laser printing would not be very practical. At this point, the plastic toner is just sitting on the paper. If you touch it, the toner will wipe right off the page with no effort at all. The final step is to pass the paper past a hot wire known as a fuser. The fuser melts the plastic, causing it to bond to the paper.

Resist Transfer

The plastic in the toner will resist the chemicals used to etch the copper. So, the question is how can you transfer the toner to a copper board? The answer is simple, but it does require some experimentation. Just as the fuser melts the plastic, you can remelt the plastic off the paper and cause it to rebond with the copper board. Several elements all have to be perfect for this method to provide good results:

- **Toner** Some toners work better than others.
- **Paper** Paper that won't hold too tightly to the toner is essential.
- **Heat** You have to produce enough heat to remelt the toner.
- **Pressure** Pressure is essential to force the toner to rebond to the copper board.
- **Removal** You must remove the paper without ripping the resist from the board.

I can give you some very specific instructions and advice on most parts of this process. However, some of the steps will require a bit of experimentation to account for your particular laser printer and toner cartridge.

Preparation

Preparing the board properly is crucial to success. The board must be extremely clean so the toner can adhere to the board, and so that the etch will attack the remaining copper properly. You should cut the board to size before you do anything else. If you use any sort of saw, be sure to wear a filter mask since fiberglass dust can be hazardous. Many people find a guillotine-style paper cutter to shear boards. This works very well and produces very little dust.

When you buy copper-clad board for this method, be sure it is not photosensitive. If it is, you'll have to strip the photoresist from the board since it is not necessary. You'll pay more for photosensitive boards, so save your money and get ordinary boards. If you are making single-sided boards, it is easiest to get copper-clad board on one side only.

However, if you want to use two-sided stock, it will work, but you will simply etch all the copper from one side of the board during the etching step.

I prefer to use ScotchBrite pads (these are similar to steel wool pads, but they are finer and nonmetallic) with a mild dishwashing soap to clean the boards. You want to really scrub until the board is shiny and even a little pink in color. Some people like to use toothpaste or rubbing compound. Steel wool leaves little grooves in the copper that can cause trouble later, so I avoid it. When using any abrasive, rub in a circular motion so you randomize any grooves you do make.

Once the board is pink and shiny, you can rinse it down with water. Then use alcohol to displace the water (alcohol will dry quickly). The important part is this: once the board is clean, don't touch the surface. Oil in your skin will form a film on the board surface and this is what you are trying to avoid!

Let the board dry before doing anything else with it. Either handle it carefully by the edges or use gloves (which you'll need when you start handling chemicals anyway).

Toner and Printer Tips

Different printers will give different results. I've had the best success with an old HP LaserJet IIIP. If you don't have access to a laser printer, you might try printing your artwork on an inkjet printer (using any ordinary paper) and then using a photocopier to transfer to the special paper. Most modern copiers use the same exact mechanism that a laser printer uses. One thing to watch for though: Some copiers don't precisely reproduce things at a 1:1 ratio. For a letter, a bit of enlargement or shrinkage isn't harmful. When you have artwork with .05-inch spacing, just a little change can make a big difference.

Don't try transferring inkjet ink to the board for use as a resist. Most inkjets use water-soluble ink. They make special paper for the transfer of inkjet images to clothing, but these work by trapping the ink in a plastic substrate, and then the entire plastic sheet adheres to the cloth. The etch will wash away the ink if unprotected, and if the ink is protected, then the entire board will be covered in plastic.

If your printer allows you to set the fuser temperature, set it as low as possible. This will keep the toner from adhering too tightly to the paper. Many printers don't offer this option (the HP LaserJet 4000 series is one that I know does offer a fuser temperature setting). I've heard of some adventurous souls actually arranging to disconnect the fuser (temporarily,

I hope). This would be ideal since the toner would easily fall off the paper (although that makes it hard to handle). However, I wouldn't recommend such extreme measures—you can get perfectly good results without performing surgery on your printer (and, of course, voiding the warranty).

High resolution is not a necessity—even the oldest laser printers will do 300 *dots per inch* (DPI), which should be fine for most boards. If you have features down to 0.05 inches, you might want to try at least 600 DPI (30 dots are used in 0.05 inches at 600 DPI). If you are using heavier paper (such as transparencies), you'll want to look for a printer with a straight paper path. Many printers can output to an alternate bin (for example, in the back) to keep the paper from twisting through the printer so often. For example, an HP LaserJet 4000, when fed from the manual tray and directed to output to the rear bin, has an almost perfectly straight paper path. For many papers, though, the path won't be critical.

One last printer tip: Turn the toner darkness as high as possible. The more toner on the paper, the more likely some of it will release onto the PCB material. Of course, if you turn it up too high, you might get spots where you aren't supposed to have toner. That, naturally, would be too high.

Paper Selection

Selecting a paper is where personal preference is the major factor. The following are several options:

- Specialty papers made to release toner
- Clay-coated "glossy" paper or inkjet "photo" paper
- Label-backing paper
- Transparencies

Everyone has a preference. The specialty papers work well, but they are relatively expensive. I personally use the sheets that carry laser printer labels or matte finish photo paper made for inkjets (some people prefer glossy paper). If you want to try labels, just buy a cheap box of labels, remove the labels, and print on the glossy area where the labels used to be.

I also use transparencies on occasion. The nice thing about the transparency is that it removes very easily. You can actually peel the transparency up a bit, and if the transfer is not complete, put it right back in the same spot and keep working. You can't do that with paper, since removing the paper usually destroys it.

A word of caution about transparencies or any other paper you put in your printer: Be sure the paper is made for a laser printer or copier. That hot fuser I mentioned earlier will do a great job of melting ordinary transparencies and that will ruin your printer. So when experimenting, be careful about what you put through your printer.

You have to print your artwork in a mirror image so it will transfer correctly. This is easy to do with Eagle's *Computer-Aided Manufacturing* (CAM) processor. I usually print a copy of the artwork on a regular piece of paper. Then I'll take my special paper and cut a piece just a little larger than the artwork. You can tape the special paper down over the artwork using paper tape (don't use plastic tapes like Scotch tape—it might melt in your printer). Then you can print the page again, and the artwork will appear right on the special paper. You do need to figure out how to feed the paper to make sure the correct side gets printed. If your printer's feeder isn't marked (many of them are) you can draw an X on a sheet of paper and feed it to see which side is printed.

Be careful not to print on the back side of the paper you will use in the final process. If the hot iron touches toner, you will transfer that toner to the iron, which makes a mess. Be sure the final artwork is blank on the back.

Heat and Pressure

For most people, heat and pressure come in the form of a household iron. You can expect to make a few attempts before you get the exact heat and pressure correct. The good news is that if your results aren't good, you can simply strip the toner off the board and start over, so be patient. Once you get the process down, you should be able to get a good transfer on the first try—most of the time, at least.

One common mistake people make with the iron is setting it to as high a temperature as possible. Typically, you want a temperature just over 50 percent of the full temperature, although you'll have to experiment to find just the right temperature. Hotter is not necessarily better.

The iron does not need to be sophisticated. In fact, you may want to buy a cheap iron just for making PCBs. Ideally, the iron will have a smooth, flat bottom and no steam or other fancy features. If you must use a steam iron, make sure it is dry and the steam is set off. Also, be aware of the holes on the bottom of the iron for steam. This will cause uneven pressure, so try to use the area of the iron that is flat. You'll find that a basic iron is quite inexpensive, so it is best to just avoid steam irons altogether.

In theory, anything that will get hot, that has a flat surface, and that can apply pressure to the board will work. If you have access to a commercial ironing press, for example, you might try that. Some people use laminators (in fact, Pulsar—formerly Dynart—sells a laminator specifically for this purpose). However, for relatively small boards an ordinary iron will work fine.

When you are working with coated inkjet paper, you can tell when the pattern has transferred. For some reason, the image becomes very visible on the reverse side of the paper when the transfer is complete. You can often tell which parts of the board require more ironing, but this depends on the paper you use.

One safety note: Copper is a great heat conductor. After ironing, the board will be very hot. Be careful not to burn yourself or scorch your work area.

Removal

Depending on your paper, you may find it difficult to remove the paper. The toner will fuse to the board but remain bonded to the paper as well. One advantage to using transparency paper is that it will usually just peel up from the board with little difficulty. For paper, you should let the board cool somewhat and then place the board in water. For most paper, you have to resist the strong urge to tug at the paper as this will tend to also remove the toner from the board. The water will swell the paper and eventually, the board will float free. Some papers may need a gentle rubbing to get the paper off of the board, but try to let it separate of its own volition first.

It isn't unusual for paper fibers to remain in the toner. As long as the fibers are completely on the toner areas, that isn't a problem. However, if the fibers are over otherwise bare copper, they could interfere with etching. If this is the case, wait for the board to cool and scrub it with a toothbrush or a wire brush. It shouldn't take much pressure to remove the paper fibers and once the board is cool, the toner will be affixed pretty well.

The board may not be perfect when you are done. For small imperfections, you can use a permanent marker (like a Sanford Sharpie) to touch up voids in the resist or connect an errant trace. If the board is too bad to correct, you can strip the board with acetone and start again. Acetone is inexpensive at the hardware store. In a pinch, you can use nail polish remover, which is usually just acetone and fragrance (with a much higher price tag).

Patience is a virtue when it comes to removing the paper or other media. This is the easiest place to ruin the board. The exact method that works best will vary depending on what type of paper you are using.

████ ████ # Etching

Once you have the toner on the board (and any corrections made with a marker), you are ready to etch it. Even if you want to use the photographic method discussed in Chapter 9, you'll still use the same techniques.

This step requires you to use some chemicals. These chemicals aren't terribly dangerous, but you do need to take some simple precautions. Be sure to read the material safety data sheets for the chemicals and take common-sense precautions (see Chapter 9 for more information about safety in general and material safety data sheets in particular). Wear gloves, don't wear good clothes, work in a well-ventilated area, and don't reuse containers that have contained chemicals for other purposes.

Three etching chemicals are commonly available. Ferric chloride is cheap and easy to get (RadioShack even sells it). It is a nasty, dark brown fluid and it will stain everything it touches. You have to be very careful with this stuff. It will stain your hands, your clothes, floors—just about anything. I once thought I had cleaned out a container that had ferric chloride in it and rinsed it out one last time on my back patio. That was probably six years ago and the stain is just now getting to the point where you have to look for it to find it. Ferric chloride is relatively safe and has a long shelf life. In most places, you can simply pour it down the drain when you are done with it (check with local authorities first).

The other choice is ammonium persulfate. This chemical is usually sold in crystalline form. You add water to it and then use it as an etching fluid. You don't have to worry about staining, but once you mix the chemical it has a definite shelf life. In addition, ammonium persulfate will eat through permanent marker ink. Sodium persulfate is sometimes used instead of ammonium persulfate because it won't attack permanent marker ink. Sodium persulfate is also somewhat faster than ammonium persulfate, but it is also more expensive and you should use it quickly after it is mixed. The persulfate chemicals are clear when fresh (you can see what you are doing) and turn blue when they are spent.

You can also get ferric chloride as crystals that you mix with water. The advantage to this is that most freight carriers will carry the crystals with no problem. Be prepared to pay for hazardous material charges if you order liquid ferric chloride. In the United States, the post office won't even carry it. You have to use an alternate carrier like the *United Parcel Service* (UPS).

Ferric chloride works best when it is hot (about 100 degrees Fahrenheit). The persulfate chemicals can also be heated. Both etching chemicals also work better if the boards (or the etch) are agitated to make sure fresh chemicals reach the board. This can range from rocking the container to using an air pump to make bubbles in the solution. Don't use an air stone, though. A plastic tube with pin holes will work and the etching chemicals won't destroy the plastic.

Speaking of plastic, be careful about the type of container you use for holding the etching chemicals. Plastic works well and, in fact, Tupperware or Rubbermaid containers are perfect since they are resistant to the chemicals and you can seal them against spills. You can also use glass, but never use a metal container. Some people seal the board with the etchant in a heavy-duty, sealable plastic bag (like a Ziploc bag). This is ideal with the short shelf-life persulfate chemicals. You can put the dry mix in the bag, and then add the board and water. Since the shelf life is not very good, you can simply dispose of the bag and the chemicals when you are through. Just be sure to use a heavy-duty bag that has a positive seal.

Agitation is not absolutely necessary with any of the chemicals. However, it does speed the process considerably and provides more uniform results. You don't actually need heat (unless using sodium persulfate). At room temperature, though, the etching process can take a very long time.

You can agitate the solution by rocking the container or using an aquarium pump. A tube with holes along the length will distribute bubbles through the solution, and that works very well. Heating can also be accomplished with aquarium equipment—namely an enclosed heater. Be careful not to overheat ferric chloride as it will give off toxic fumes. You should work in a well-ventilated area anyway. Keep the solution around 100 to 110 degrees Fahrenheit. You can also work on a hot plate (on a low setting) or use infrared heat lamps. I have a friend who puts his etchant in an old microwave used just for this purpose, but I still prefer an aquarium heater.

You can purchase specially made tanks to etch boards. These tanks are often nothing more than a Tupperware container, an aquarium pump, and an aquarium heater. You can get more expensive tanks that actually spray the etchant. You can also create your own tank. Figure 8-1 shows a commercial etching tank. Usually, a plastic tray on the inside of the tank has an air hose with holes in it, as well as a groove to actually hold the board.

If all this seems overwhelming, don't worry. Just get some ferric chloride, a plastic tray (or a container or a heavy bag), and try etching boards at room temperature. If you are really adventurous, add a little heat to the process. Once you get the feel for making boards, you can experiment with different chemicals and methods. The only downside to etching at room temperature is that it takes a long time. Not only is this boring, but it also can cause

Figure 8-1
A commercial etching tank (courtesy of MG Chemicals)

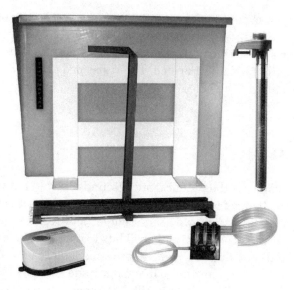

etchant to undercut fine traces, so make sure you aren't making very finely detailed boards right away.

You can use ferric chloride repeatedly, although it will eventually wear out and you'll notice that the boards simply aren't etching (or they are etching very slowly). The persulfate chemicals won't last long after you mix them, but if you use them on too many boards, the solution will turn blue.

Some people will add chemicals to the etch to extend its life or speed the etching process. My advice: Never add anything to these chemicals! Although these reagents are relatively safe, if you start mixing things you can produce some seriously poisonous gas. The etchant chemicals are not so expensive that it is worth the risk. Just buy more etchant.

Once the board is out of the etching bath and the copper you don't want is gone, you can rinse the board in water. If you are using ferric chloride, be careful. Just a little—even heavily diluted in water—will still stain almost anything it touches. It may seem odd, but I rinse my boards outside over the grass. You have to make sure the boards are absolutely free of etchant before you handle them with your bare hand or let them touch anything that might stain.

You should inspect the board to make sure all the copper is gone. If it isn't, you can dry the board and put it back in the etch. You can also scrape away small bits of copper with a hobby knife or other sharp instrument.

The final chemical step is to remove the resist coating. Again, acetone and elbow grease will get the resist off. You can scrub with a Scotchbrite pad —you can't harm the copper that is left. However, I usually put this step off until after the holes are drilled. The resist pads make it easy to see where you have to drill the holes.

Drilling

Oddly enough, all the chemical steps sound difficult, but they are easy once you get the hang of it. Drilling holes sounds easy, but it is by far the hardest part of fabricating your own boards. Forget your regular hand drill. The problem is that you need very small drill bits to make the holes. The smallest bit you can typically find in a home improvement store is 1/16 of an inch, which is way too big. For many holes in a homemade board, a 1/32-inch bit will do. A #66 numbered drill is better still. In a pinch, you can get a 1/32-inch bit in a set of drill bits sold for use with Dremel tools. However, these bits are normal steel and will dull quickly against the fiberglass PCB material.

Ideally, you want to use a carbide drill bit. These are available from a variety of sources (often resharpened bits used by large manufacturers). However, these bits are brittle and require high-speed rotation. A Dremel or similar tool isn't really fast enough, but it will usually do the job. However, if you try to use the Dremel by hand you will snap the bit as soon as you move off the vertical (be sure to wear eye protection). Dremel makes a drill press attachment for their tools that works very well and will quickly pay for itself in preserved drill bits.

Another alternative, if you have access to one, is a jeweler's press. This is a tiny precision drill press that works at a higher speed than a normal drill press. Normal drill presses are typically not suitable for this kind of work. They are far too slow and they usually have an unacceptable amount of run out (that is, the drill does not rotate exactly on its center). Some people do use a regular drill press, though, but the results won't be as good as using a Dremel or a press made for precision work.

The other problem with drilling is successfully aligning the holes. Resistors, capacitors, and other simple components are no problem —if the holes don't line up exactly, it won't be a problem (although it will look ugly). Integrated circuits and some other components (like some switches or other components with rigid leads) require precise spacing of leads.

If you have a well-equipped machine shop, you can mount the board in a cross slide vise (or even an indexing table) and use that to exactly space your holes. Most of us have to find a simpler method. One trick is to take a piece of perforated board (the kind used to build prototype circuits) and line it up with the board. You can immobilize your board with the perforated board using paper clamps, masking tape, or rubber bands. Then you can drill through the perforated board into your board and your holes will wind up nearly perfect if you get the initial alignment correct. This takes some practice but provides very good results once you get the hang of it.

Eagle's pad patterns have a hole in the middle. If you leave the resist on the board, this will form natural depressions that will help you guide your drilling. You can also use a center punch to make a dimple where you want the hole. It only takes a minute to use a spring-loaded center punch to indent each hole and this also helps to guide the drill bit.

One advantage to using surface mount is that you have fewer holes to drill. Once you've drilled a few PCBs by hand you may decide to switch to surface-mount technology! I've even seen people make quasi-surface mount boards by making a normal PCB, not drilling any holes, and simply soldering leads directly to the copper pads. Of course, through-hole integrated circuits require some lead bending so they sit flat. I'd suggest just drilling the holes, but this will give you an idea of how far people will go to avoid the laborious drilling step.

Finishing Steps

Once the board is drilled, you can remove the resist (if you haven't already done so). With time, the copper on the board will corrode. To prevent this, you can either coat the entire board with solder or tin plate it. Conventional tin plating is out of reach for most of us. However, several "electroless" tin plating solutions are available. Liquid Tin is very easy to use, works at room temperature, and produces a good tin coating. It is apparently a nasty chemical, though, so be sure to read the material safety data sheets for it before using it. Another popular product, TinIt, provides similar results, but it requires heating. The tin plating is attractive, and improves the solderability of the board.

If you don't want to tin plate the board, you should solder it as soon as possible. If you store the board for a while, you may have to clean the board (with Scotchbrite) again. After you complete the board and everything is working, you may want to coat the copper to prevent further corrosion. Several coatings, such as silicon lacquer, are used for this purpose.

Step by Step

Creating a board consists of a lot of steps. Here's a checklist for your review:

■ Print your artwork, mirror imaged, on a regular piece of paper. Use the darkest setting, the lowest fuser temperature (if possible), and the straightest paper path.

■ Cut out a piece of your selected transfer paper to cover the artwork.

■ Use paper tape to affix the transfer paper to the regular paper over the printed artwork.

■ Print the artwork again to put the image on the transfer paper.

■ Cut the blank board to size and scrub the PCB until it is shiny and pink. Make sure the board is dry and oil-free.

■ Place the transfer paper face down on the board and use a clothes iron to "iron" the artwork on the board. The heat, pressure, and time required will require some experimentation.

■ Remove the transfer paper. For most paper, you should soak the board in water until the paper floats away from the board. Some papers work better if you peel them away while warm (in particular, transparency film).

■ Touch up any problems with the board using a permanent marker.

■ Place the board in your preferred etching solution. Heat and agitation will speed up the process. Remove the board when all the unwanted copper is gone.

■ Wash the board clean.

■ Drill the holes in the board using a Dremel drill press and carbide drill bit.

■ Strip the resist from the board using acetone.

■ If desired, coat the copper with a solder or electroless tin plating solution.

Double-Sided Boards

So far I've only mentioned single-sided boards. Making double-sided boards with this technique is elusive. The problem is getting the artwork on both sides of the board to line up. You can try using oversized pieces of transfer paper for both sides. Carefully line the artwork up (holding it up to a lit window or a light table is helpful) and use tape to seal three sides of the paper together. Insert the blank board into the fourth side and line everything up.

You might want to drill a few holes to see how the holes line up. Then—if you are very, very careful—you might be able to iron the artwork on both sides closely enough to match.

Honestly, I haven't had a lot of success making two-sided boards using this method. It is tough to keep both sheets of artwork and the board in the same positions throughout the entire process. Even a small amount of misalignment will ruin the board. If you are set on making double-sided boards, you should probably try the photographic method in the next chapter (or send them out to be made).

No reasonable way exists for making plated through holes for components and vias. If you make your own two-sided boards using any process, you'll have to solder components on both sides for components that have connections on both sides. In addition, you'll have to insert a bare wire through each via (don't forget to drill the vias) and solder the wire on each side. You can then cut the wire flush on each side.

Some companies produce eyelets that you can fit into holes to provide the illusion of a plated through hole (these are often used to repair PCBs). The eyelets are inexpensive, but you also need a special tool that is a larger (but one-time) expense. I personally just use a wire soldered to both sides of the board.

In Summary

It seems like a lot of work to make a board by hand, and it is. However, with a little practice, it is possible to create nice-looking boards on a regular basis. However, these boards are no match for commercially made boards. Proper solder masks, for example, cannot be reasonably created in a simple setting (they do make pens used to repair solder mask that you may be able to use to apply mask to small boards, but it is tedious and difficult on fine boards). Silk screening is also rarely seen in homemade boards (although you can use iron-on toner to make a black component marking). Double-sided boards are difficult to make with this technique (although the photographic method in the next chapter can produce acceptable two-sided boards), and no way to create plated through holes exists.

Still, for one-sided prototypes the laser printer method is hard to beat for quick results. You can lay out a board in the morning and have a finished board in the afternoon. I personally find the laser method easier to do than the photographic method, although certain people feel the opposite is true. You can read the next chapter and decide for yourself.

Photographic Boards

It is hard to imagine a world without photography. Yet Nicéphore Niépce took the first picture (which took eight hours to expose and had to be viewed at a certain angle) in 1827. Prior to then, making a semipermanent record of something you could see was the realm of artists and science fiction writers (like the author Charles François Tiphaigne de la Roche, who predicted photography in his 1761 book, *Giphantia*). Niépce later teamed with Louis Daguerre who, after Niépce's death, produced the now familiar daguerreotype process for photography.

Needless to say, photography has come a long way since then. Color photography is commonplace. Three-dimensional photography isn't common, but it is available. Edwin Land's wonder—the Polaroid instant picture—is no longer a marvel and is now in danger of extinction thanks to digital photography.

When you think of photography, you probably think of the traditional things: vacation snapshots, wedding pictures, or perhaps a photograph in the newspaper or on a web site. However, photography plays an important role in many areas you might not expect. For example, until recently, nearly all modern printing was done using a photographic process. Although computer-generated printing is quickly gaining ground, a lot of printing is still done with photographic methods. The integrated circuits that make your PC, your cell phone, and all the other electronic wonders we take for granted would be impractical without a form of photography. Technically called photolithography, the process involves making huge negatives (or, sometimes, positives) known as photomasks. Just as the local Wal Mart uses a light beam to enlarge your small negatives into larger prints, semiconductor makers use a light beam (or an X-ray beam) to reduce a huge photomask to a tiny integrated circuit pattern. Photosensitive chemicals on the silicon wafer react to the light (or absence of light) and provide a resist that prevents a subsequent chemical bath from removing certain parts of the chip while eating away the unwanted material.

Does that sound familiar? Making a *printed circuit board* (PCB) involves using a chemical bath to remove copper from areas where you don't want a conductor. You also have to protect the copper that you do want so the chemical doesn't remove it. Although the photolithography process sounds high-tech, it is actually well within the reach of the home or small lab.

With modern computer printers, making photomasks for an ordinary PCB larger than the actual board is unnecessary. When making photomasks by hand (or making them for microscopic *integrated circuits* [ICs]), making them several times their actual size is an advantage because any error gets divided by the scale factor. That is, a 1-millimeter error on a 10X photomask will only result in a 0.1-millimeter error on the finished product. But modern printers can produce same-size photomasks that have no appreciable error at the tolerances you are likely to require.

Materials Needed

Making boards via photolithography requires a bit more set up than the laser printer you read about in Chapter 8, "Boards from a Laser Printer." However, the results—once you get the hang of it—are more consistent. Once you get the resist on the board, by the way, the process is exactly the same as the laser printer method. I won't be talking about etching and drilling the board in the chapter. If you skipped Chapter 8 because you wanted to do photolithography, you'll just have to read the second half of it after you finish this chapter.

The basic elements you'll need include the following items:

- A copper board with a photosensitive coating
- Transparency film
- An exposure lamp
- A contact frame
- A foam paint brush
- Some plastic trays or containers
- Developer
- Resist stripper (optional in some cases)

It is possible to buy photoresist in a bottle and apply it to blank boards. However, it is difficult to get an even coating, so most people just buy the boards already coated. Of course, this is somewhat more expensive than buying blank boards. Also, it makes it more difficult to cut the boards since cutting may peel off the photoresist layer. If possible, make the board first and then cut it after etching (but before drilling). That way, no resist is left to damage, but you do wind up wasting a bit of the board.

If you use precoated boards, or you coat your own, you need to make sure you get a compatible developer and stripper. For example, negative photoresist boards use either a dry film resist or a liquid resist. The dry film resist is very tough—it is much harder to accidentally damage the resist in handling. Also, the developer for these boards is not solvent-based and is easier to work with. On the other hand, the liquid resist is fragile but is better-suited to artwork with thin traces (15 mils or less). The developer is solvent-based and not as easy to handle as the dry film-compatible solution. MG Chemicals, a popular vendor of presensitized boards, makes positive boards that are easy to work with under normal lighting (the negative boards usually require a yellow "bug" light). However, they take yet another type of developer, so be sure you get compatible boards and chemicals.

Printing Artwork

As you'd expect, the first step is to print your artwork on a piece of transparency film. If you are using a laser printer, make sure you are using transparency film specifically made for a laser printer (since some film will melt inside the printer). If you are using a different kind of printer, you should probably print on regular paper and then use a photocopier to transfer the image to a transparency. The key is that you want the black area on the transparency to be very dark. You also want to print at 600 dots per inch or higher, which nearly all modern printers can do. Even the oldest laser printers can achieve 300 dots per inch, and that is adequate for most boards (if you expect to get .05 inch features, that's 15 dots at 300 dots per inch—you wouldn't want to go much below that, though, without more dots per inch).

You can get photoresist (or boards coated with photoresist) that requires a positive or a negative image (although the major U. S. supplier of negative photoresist boards, Kepro, ceased operations about the time this book was written). That means you must print your artwork so that the black areas represent the copper if you are using a positive board, or the inverse if you are using a negative board. What's more, since you have it on a transparency, you'd think you would not need to print a mirror image. However, you should print a mirror image (easy to do with Eagle—in fact, some layers are mirrored automatically when you print). With a mirror image, you can place the printed side directly into contact with the board, which is good. If you don't mirror the image, you'll have the transparency between the board and the image, which will cause small errors in your edges. Still, these errors are probably acceptable for most boards you'd try to make yourself.

If you have a modern laser printer, you can probably just print a single transparency and be done with it. If you have an older printer that isn't very dark, you can try printing the transparency more than once, or printing two transparencies and sandwiching them together. Again, though, having a layer of plastic between the images will cause slight errors, but probably not enough to notice unless you are trying to produce very exacting boards. Still, it is easy enough to produce mirrored artwork with Eagle's *Computer-Aided Manufacturing* (CAM) processor, so you might as well take advantage of the software and produce the image on the side that will contact the board.

You can also take your artwork (on plain paper) and have it transferred to film by most print shops (or check with the printing department of your local newspaper). You may also be able to find a print shop (or PCB maker) in your area willing to do a photoplot for you. Photoplotting takes special files (you can create them with Eagle; see Chapter 10, "Outsourcing Boards") and uses a machine to produce a film of your artwork. If you are asking a print shop to make a film from your printed artwork, they will probably want a reverse of what you need. That is, if you need a negative, you'll supply them positive artwork. If you want a positive, you'll supply the print shop a negative. They can then make a 1:1 print on Kodalith Ortho (which is a negative film) emulsion side down. The cost of this varies widely, although if you can make friends with the local school newspaper staff, you may be able to get it done for the cost of the film.

The camera method, by the way, used to be the standard way to go. Special reversing films are also used to produce negative images. However, with computer output devices having 600 dot per inch resolution (and higher), these really do not make much sense anymore. Almost everyone can at least get access to a good-quality laser printer and do the work themselves. I have seen elaborate setups that people use to shoot artwork on a regular camera and then transfer it to film, but again, with computers, scanners, and laser printers almost universally available, I just don't see the point anymore.

Whatever method you use, the key is to get a solid dark pattern. If the dark areas have pinholes, you will get pinholes in the finished product as well. If you aren't getting satisfactory results, try a darker setting, a new toner cartridge, or a different printer. Different transparency media can help too. Some people prefer to print on vellum, which is a semitransparent paper, but this will increase the exposure time since it is not as transparent as the transparency film. If you have just small imperfections, you can touch them up with a black marker on the film.

If you don't personally own a laser printer, but you do have access to one at work, a friend's house, or school, here's a tip: Install the correct printer driver for the printer on your computer. You can download current drivers from the vendor's web site and it won't actually check to see if you have the printer (at least, in most cases). Then you can direct the printer output to a file instead of a physical printer. Copy the file to a floppy (or a CD-ROM, or e-mail it to yourself at work) and then you can simply copy the file to the printer. For example, in Windows you can open a command prompt and enter the following:

```
COPY output.prn LPT1 /B
```

This is assuming, of course, that output.prn is the output file name and the printer is on LPT1. Eagle's CAM processor, of course, can already output to any of the devices it knows about using a file, so in that case you don't even need the drivers (see Chapter 7, "Eagle Output," for more about the CAM processor).

Exposure

If you have film and a sensitized board, you are ready to start. If you want to sensitize your own boards, you can buy bottles of photoresist. It is easy enough to apply on the board, but you want a uniform coat. Also, you'd like to apply the same amount each time, so that exposure and development times will be predictable. You'll want to clean the board until it is shiny (use a ScotchBrite pad) and wash any oils off with alcohol. Then you can use a dust-free cloth (makeup removal pads work well) to apply the photoresist. Most photoresist is not active until it dries, so you can handle it in a lit room (although check the instructions for the actual photoresist you are using). You want to spread the photoresist uniformly in all directions (or spray it evenly if it is in an aerosol can; aerosol cans are hard to find these days, doubtlessly because of environmental concerns). When you are done, put the board in an area with no fluorescent lights (unless your photoresist specifies differently). Place the wet side facing down so that you don't get dust settling on the photoresist emulsion. You may have to dry the board overnight (or dry it in an oven).

Precoated boards are not much more expensive than blank copper boards. The precoated boards will have an even coating of resist, so I haven't tried coating my boards myself in a long time. It just isn't worth the extra hassle to save a few pennies. Especially when the results are usually better with the precoated boards.

Most modern positive photoresist—either preapplied or that you apply yourself—is sensitive to ultraviolet light. That means that sunlight and fluorescent lights will expose it, but ordinary incandescent lights won't. You probably don't want to leave coated boards out in the light for a long time, but short exposure to nonfluorescent room lighting won't do the board any harm. The negative boards usually specify that you work with them under a yellow light (a bug light will work). The presensitized boards are in an opaque bag and you should not open the bag until you are ready to use the board. Be careful to handle the board only by the edges. If you damage the resist coating, you'll ruin the board.

You can purchase contact frames (at a photo supply store) that make a sandwich of the board, the artwork, and a glass weight. The glass presses against the artwork and prevents light from leaking under the traces (which is also why it is best to have the artwork printed on the surface that will contact the board). Of course, you can simply put the board on a flat surface, lay the artwork on it (make sure you have it oriented correctly, and not flipped—it should look like the finished board, not a mirror image). Then place a piece of glass from a picture frame (3-millimeter thick glass is good) over the entire assembly. You can use some transparent tape to hold everything together so it doesn't move. You can also get expensive contact frames that use a vacuum to lock everything in place. This is essential for fine resolutions, but for many homemade boards, it is overkill.

The boards (or photoresist) you use should give you some guidelines regarding exposure time with a fluorescent (or other) light source. Of course, lights vary, but if you place the light about 5 inches from the artwork, you should expose the board for about 10 minutes.

Some people prefer to use photoflood bulbs for exposure. I know a few people who put their contact frames on the windowsill on sunny days. You may ruin a few boards before you find the exact timing required for your setup. However, the exposure times are not very critical, so as long as you give the board enough light, you should be able to salvage the board. The only problem arises when the transparency is not completely dark. Then overexposing will cause some (or all) of the resist you want to keep to wash away during development.

If you want to be scientific, you can make a test strip with your new setup (although it will cost you a board). Just set up as usual for exposure, but before you turn on the lamp, cover roughly 90 percent of the board with a piece of opaque material like cardboard. Every 2 minutes, move the cardboard so that about 10 percent more of the board is exposed. After 20 minutes, you can develop the board (and, perhaps, even etch it). The original 10 percent area was exposed for 20 minutes and will look good if your artwork is opaque. The next area will show you the exposure for 18 minutes, and so on.

If you are trying to get fine detail on your boards, dust is your enemy. Make sure you are scrupulously clean when handling the film. The contact glass must be dust-free. You may want to buy lint-free gloves (from the photo store) to wear while working with film.

Developing

Just like a photographic negative, the exposed board has to be developed. The developers used are often very corrosive (or dangerous solvents), so rubber gloves and eye protection are essential. For positive boards, the developers are essentially sodium hydroxide (or something that forms sodium hydroxide when mixed with water), which is pretty nasty stuff—it will eat through just about any metal given enough time and heat. Liquid negative emulsions use xylene or methylene chloride as a developer. These are solvents (sometimes used as paint thinner), have dangerous fumes, and are flammable. The dry film negative emulsions develop in sodium carbonate, which is relatively harmless compared to most of the other chemicals

You need to get developer suitable for the boards you are using. Be sure to read the material safety data sheets before you start working with it and follow all safety guidelines (see the sidebar "Safety and MSDSs"). For example, if the developer is a powder, you should follow the mixing directions exactly. If the developer is too strong, it will remove all of the photoresist. Weaker developer is better than strong developer!

You'll need to transfer the board to a plastic tray full of developer. Some developers need heat, but others work at room temperature. Consult the bottle to find out what the vendor recommends. You'll also have to read the directions to find out if (and how much) water you have to add to the developer. Be careful not to touch the resist at this point because it is still fragile. The directions for your developer may tell you to wipe the board with a foam brush or sponge. Or it may just tell you rock the tray for a few minutes and rinse the board in water.

At the end, all the resist that was not exposed should be gone. The remaining resist should be hard and can be safely handled.

Finishing

Once you have the resist on the board, you can do the classic etching and drilling steps explained in Chapter 8. There's really no difference at this point. The etch step removes the copper where the resist is not present. If

Safety and MSDSs

The processes mentioned in this chapter require several chemicals. Some of them are rather nasty and you should be prepared to exercise the proper care when handling and disposing of them. Years ago, when you bought chemicals you were expected to know what to do (and not do) with them. Today the seller is responsible for making *material safety data sheets* (MSDSs) available to you.

You can ask your vendor for an MSDS on any chemical you purchase, although it is usually easier to find the data online at their web site. In addition, several web sites attempt to aggregate MSDSs from around the Web so you can search for any chemical you need. For example, www.ilpi.com/msds/ is a good source of links. Of course, a Google (or other) search for the chemical name and MSDS will usually turn up what you need, too.

Take the time to read the MSDS before you use any chemical. You'll find the same information (required by law) on all MSDSs, but the format is not standardized. The topics include the following:

- Definitions and acronyms
- Chemical product and manufacture identification
- Composition
- Physical data
- Fire fighting
- Hazardous identification and first aid
- Stability and reactivity
- Accidental release measures
- Handling and storage
- Exposure controls and personal protection
- Toxicological information
- Ecological information
- Disposal considerations
- Transport information
- Regulatory information

(continued)

Safety and MSDSs *(continued)*

You want to pay special attention to the fire, first aid, toxicology, and disposal data. You'll find that an astounding number of chemicals are suspected (or proven) to cause cancer. Don't improperly dispose of chemicals! Doing so could be a criminal offense and subject you to a fine or worse.

If you are storing chemicals, write down the poison control center's phone number and affix it to the container (or nearby). Be careful not to store chemicals where they might leak, spill, or be accessed by children, pets, or anyone for that matter.

A few commonsense precautions:

- Always work with good ventilation. If you work in your garage, open the garage door. If it is too cold to open the door, work another day.
- Make sure running water is available nearby.
- Never work without a telephone. It is even better if you don't work alone.
- If you are working with flammable chemicals, be sure to have the proper type of fire extinguisher. Not all fire extinguishers are suitable for all chemical fires.
- Have paper towels to absorb spills.

Of course, you should also have all recommended safety equipment —goggles, gloves, and an apron, for example. Ferric chloride, since it stains so easily, is a great teacher. You can be certain you are not getting any on your hands or the surrounding area, and you will still be amazed at how many little droplets get on your gloves, apron, and work area when you are done.

you have imperfections in the resist pattern, you can touch them up with a resist pen or a permanent marker (see Chapter 8 for more details).

You can get a stripper that will remove the resist after etching. However, the positive resists will not hinder the solderability (the heat burns the resist away), and the resist serves to protect the board from oxidation. If you don't remove the resist, you won't have to protect the copper against corrosion. However, if you are planning on tin plating, you'll probably have better results if you strip the resist after the etch step. If you are using a negative resist, check the vendor's recommendation.

Since you probably didn't cut the precoated blank boards to size, you can cut them after the board is complete. I like to cut them after etching, but before drilling. That way if I have an accident while cutting the board, I don't waste time drilling it. When cutting fiberglass with a saw, be sure to wear a dust mask, as fiberglass dust is not good to breathe. You can read more about cutting boards in Chapter 8.

Two-Sided Boards

Making two-sided boards is a bit of a challenge, and you should get used to making single-sided boards first. The obvious challenge is registration—getting the two sides to line up properly.

The easiest way to line up the two sides is to make the artwork a bit larger than the board. Line them up perfectly and tape them, temporarily, together. Make certain the two layers are registered perfectly before continuing. Then, in the area outside the board, make holes at each corner through both layers with a large needle. Now you can place one layer in the contact frame, followed by the board. Use some tape to hold everything together. Then put the other transparency on top of the board and use the holes to line up the two layers. You can put needles through the holes to act as a guide. Finally, use a bit of tape to seal everything together (then you can remove the needles). Expose one side, then flip the whole arrangement over, and expose the other side.

Everyone has a favorite method for registration, and this just happens to be mine. You can also affix the artwork to one side of the board and drill a few holes in the board at this time. These would be holes you would drill anyway, and they should be in two opposite corners of the board. Then you can drill the same holes in the other piece of artwork and use the holes to align the sandwich. I don't like to do this because it's possible to damage the photoresist while you are drilling the holes (not to mention the debris caused by the drilling).

Another technique some people use is to simply produce two single-sided boards and laminate them together. I personally don't care for this method, but it does work (although the board is twice as thick as necessary, of course). You can also cover one side of the board with paint (or tape or a label—anything that will resist the etching process) and then expose and etch the other side. Then you can strip the bulk resist and finish the other side.

Another problem with making two-sided boards is handling the undeveloped boards. With emulsion on both sides, you'll have fun trying not to ruin the photoresist. You also can't let the board touch the bottom of the developing tray because that will be enough to scrape the photoresist layer. The best strategy is to use a larger blank than required. If you leave one-half inch around the artwork in all directions, you will have an area where you can safely handle the board. Some plastic clips (be sure no metal springs are in them) can grip the handling zone, holding the board off the bottom of the tray.

These methods can be wasteful because you need a board larger than the actual artwork, and you also need a larger transparency to handle registration. However, the amount of waste is small and not usually of much importance.

Of course, all the restrictions mentioned in Chapter 8 still apply. You have to solder components on both sides and insert wires through vias, for example. No plating through your holes takes place as there would be in a professionally made board.

In Summary

Compared to laser printer PCBs, the photolithography process is complicated and requires more chemical steps. However, photolithography can produce better and more consistent results than the laser printer method.

I personally like to work with the positive photoresist boards. You can handle them under incandescent light, and it is easy to create positive artwork. The chemical required to develop the board isn't very hard to handle and works at room temperature. In the end, for most boards, it doesn't matter which process you use. Once you're set up, one method isn't much different from the other. You do need to make certain you have the proper chemicals that match the boards or photoresist that you are using. The directions in this chapter are guidelines. Always defer to the instructions provided with the board and chemicals you've selected.

Although photolithography results in better boards than toner transfer (in general), the boards are still not up to the standards you can expect from a professional board house. Your boards won't have silk screening, plated through holes, or solder mask. In addition, a board house (that doubtlessly has expensive equipment) can make finer details than you are likely to achieve in the small lab. On the other hand, you can make boards quickly and inexpensively once you overcome the initial hurdle. With a professional board maker, you'll usually pay quite a bit to have boards made in small quantities. If your board has a mistake, you'll not only lose your money investment, but you will have wasted days or weeks of waiting.

Outsourcing Boards

I once had an older house that needed plumbing work. I quickly grew tired of paying the plumber to make repairs and started doing most of the work myself. This came to an end when I put in a new hot water heater. It was fine for an hour or two, and then it started shooting water into the air like a fountain (apparently I didn't get all the PVC joints solvent-welded). After that, I realized that the plumber was cheaper than doing it yourself and then having to call the plumber to fix your work anyway!

After many years of making *printed circuit boards* (PCBs), I've almost come to the same conclusion about board making. Sure, you can do it yourself, but the results are never as nice as the results you get from a professional board maker (who has a substantial investment in expensive capital equipment). When it comes to boards, you can have them high quality, cheap, and fast—as long as you are willing to pick two of those three attributes! That is, you can get high-quality boards, but not cheaply. You can save money, but you'll have to wait (or sacrifice quality).

Before the computer age, it was troublesome to have to send your artwork away and wait for boards to arrive. In those days there was a big advantage to having a local company make your boards. You could drive over with a tube full of artwork on film. Today you simply send your files in an e-mail, they charge your credit card, and you can get boards back in a few days via FedEx. Manufacturers overseas can be very inexpensive, if you are willing to wait for shipping.

What's the downside? The real expense to board makers is the set up to make one board. All the expense is in the first board. It is not unusual to pay $100 or more just for tooling. That means that after you pay the $100, you can then get that first board for the normal price (which depends on the area of the board and a few other factors). However, some board houses specialize in making single prototype boards. They will probably cost more per board than one that charges a tooling fee, but for one or two boards, it will be cheaper than paying the tooling costs. Of course, if you are making hundreds of boards, the tooling charge amortizes to a small amount per board.

Types of Vendors

You'll encounter a few different categories of board houses.

- High-volume vendors are best suited to making large runs of boards (production quantities).

- Prototype houses specialize in making just a few boards. Some of these can also handle high-volume jobs at a fair price.

- Hobby-oriented vendors offer you free or low-cost software for board layout. The catch is that the files the software creates are proprietary to that company. You can't have boards made anywhere else (although some will create standard files for you for a fee).

- Unusual manufacturers use odd processes (such as machine milling, lamination, or organic chemical processes). These can sometimes be a good deal for a one-time board, but in general I'd avoid these companies as well.

Many people use the hobby vendors simply because they don't know how to use a PCB program, and these manufacturers offer relatively low prices with no setup fee. However, unless you are completely certain you'll never want to make even a small quantity of boards, you should avoid getting locked into these programs. Besides, I don't know of any proprietary program that is as full-featured as the free version of Eagle (although other programs may do larger boards, they don't have extensive libraries, schematic capture, and autorouting). However, these companies usually won't take any files that aren't generated by their own programs.

The milling machine process is not bad, although it usually just eats around the traces and leaves copper in blank areas where it shouldn't matter. If you can afford the $10,000 plus price tag for one of the machines, then it is probably a great deal. However, paying someone else to run your board on his or her machine isn't such a good deal. The boards won't look professional and you'll still have to wait. What's more, the boards probably won't be much cheaper than what you can get from a more conventional vendor.

I had a bad experience with a company that used an organic process to make PCBs. It feels good to help the environment, but most people make their PCBs using photolithograph and standard chemistries for a reason: They work! Although organic chemistry for PCBs may one day be the standard, I don't want to be the test case for it. If they perfect it, you can bet that nearly everyone will start using it (and save on expensive disposal costs).

You probably will also want to avoid small garage-based companies that will offer to make boards for you. Although these can be cheap, you can do the same work yourself (just read Chapter 8, "Boards from a Laser Printer," and Chapter 9, "Photographic Boards"). If the company only offers single-sided boards, for example, or can't do silk screens or solder mask, they are probably just doing the kind of things you could do yourself. If you are going to pay, pay for quality professional boards.

Preparing for Outsourcing

In theory, sending your board out for production is simple. Some vendors will accept Eagle BRD files directly. CadSoft maintains a list of those that do on their web site. In fact, CadSoft allows board makers to use the free version of the software (without limit) to produce boards for Eagle users. In other cases, you have to use the *Computer-Aided Manufacturing* (CAM) processor to produce Gerber and Excellon files (see Chapter 7, "Eagle Output"). This isn't very hard and Eagle does a good job of writing the necessary output.

In practice, a few complications exist. First, not all board vendors have the same capabilities. You have to be sure your board complies with your supplier's rules. In addition, since it is usually very expensive to have the board tooled, you will want to make sure the board is completely correct and your files will not be misinterpreted.

When you send boards out, you need to think about a few additional things. For one, the board house will probably make your boards on a standard-sized panel. It is very likely that your board will not be as large as the panel. You'd think that you want them to cut the boards apart (individual routing). In fact, most of the time you probably do want them to cut the boards. However, if you are going to build lots of copies of your board (or have a contract assembly firm build them), it is easier to handle the boards still on the panel. In this case, you can ask the PCB maker to score the boards, but not actually cut them. Then the assemblers can work with one large panel (which is especially helpful if you are using automatic insertion machines and wave soldering). When the boards are stuffed with components, you can break them apart at the score lines. This is usually not an issue when you are making one or two boards yourself.

The other item to consider is electrical testing. No process is perfect, so it is possible to have boards with defects (although with good board producers, the number of bad boards is surprisingly small). Most PCB makers have the equipment to electrically test your boards so they don't ship you any duds. The catch? You'll have to pay another setup fee to get this service.

Getting a Quote

Some small prototype board houses quote a flat price for their work. However, in most cases you will have to answer a few questions about your board and the sales agent will give you a price. Better still, many board makers now have an online form that you fill out and it automatically creates a quote for you. This is very handy when you are trying to get a quote at 3 A.M. Even better, you can experiment to see which parameters will most influence your cost.

The only problem with automated quote forms is that they frequently ask for things you may not have thought about. Usually, a human sales agent will assume certain things about your board unless you mention otherwise. The quote forms require all the information.

Here are some typical items you'll provide for a quotation:

- **Part number** This is just your part number for the board and has no real meaning to the board maker. However, if you make a lot of boards with one company, it will help you reorder parts if you provide a part number (and revision number, too).

- **Dimensions** The vendor will want to know the size of the board in inches or metric measurements. If your board isn't rectangular, you'll need to figure out the smallest rectangle that will completely cover the board.

- **Quantity** The more boards you buy, the cheaper the per-part cost. You can usually ask for pricing at several quantity levels.

- **Layers** Most board houses always run at least two layers, so one-sided boards are no cheaper than two-sided ones. Of course, if you have the full version of Eagle, you can create boards with more than two layers, too.

- **Material** Typical PCBs use FR-4 fiberglass. However, certain special boards require a ceramic substrate or other materials such as Teflon, Polyimide, or flexible materials (if you don't know if your board requires anything special, then it doesn't). For very high volumes, you can use a paper-based board (you often see this in high-volume consumer electronics). Not only are the boards cheaper, but they are punched instead of drilled (which requires a hefty setup fee, but reduces the per-board cost substantially).

- **Board thickness** You can get boards with various thicknesses. A normal board is .062 inches thick (1.5 millimeters).

- **Copper weight** A garden-variety board will use 1 ounce of copper (also known as 35 _m copper). If you are working with high current, you may need heavier copper (see Chapter 3, "Board Layout," for more about copper weight and trace widths).

- **Number of holes** The board maker will want to know how many holes need drilling in the board. This will include mounting holes and vias. The Eagle COUNT.ULP program will give you this number automatically. Special holes (for example, holes that require countersinking) will cost extra.

- **Smallest hole** Some board houses will want to know the smallest hole you need drilled on the board. Other vendors will offer you a standard rack of drill bits. Any holes not in the standard rack will cost extra. If you are using an overseas company, the drills will probably be metric. In the United States, inch sizes are more common.

- **Minimum trace width** This is just the size of the narrowest trace on the board. Most companies have an absolute minimum that they can do. Using very narrow traces will increase cost, lead time, and increase the number of bad boards.

- **Minimum spacing** This is the smallest distance between two unconnected pieces of copper on the board. Again, usually you must obey some absolute minimum. Making spacing too close will increase cost, lead time, and the number of defective boards.

- **Surface-Mount Device (SMD) side** Boards with surface-mount components may require special processing like leveling (particularly, *Hot Air Solder Leveling* or HASL). You can have SMD parts on either side of a two-sided board or on both sides.

- **SMD pitch** This is the closest spacing between two SMD pads. Some surface-mount devices have leads closer together than other devices and the board maker needs this information. Some vendors want to know the exact spacing. Others will simply want to know if you have "fine-pitch" SMDs (typically, anything 0.025 inch or below is considered fine pitch).

- **SMD pads** The number of SMD pads on the board total. For example, if your board has two *small-outline integrated circuit* (SOIC) packages with 8 pins each, the total number of pads would be 16.

- **Solder mask** You can choose if you want solder mask on either side of the board, or both sides. Solder mask increases the cost but makes it easier to assemble the boards (and is essential for automated soldering and SMDs). The most common solder mask is *Liquid Photo Imagable* (LPI). Other choices are dry film or silk-screened. Silk-screened solder mask is usually not as precise as the other methods and may not be suitable for boards with fine features. For most boards, either LPI or dry film is fine. Some companies can offer you a choice of solder mask colors.

- **Finish**   Some board makers offer various plating options. The most common is *Solder Mask Over Bare Copper* (SMOBC), which is essentially no finish. Reflowed solder finishes, tin plating, nickel plating, silver plating, and gold plating are other options. Unless you have a special requirement, SMOBC is probably the best option.

- **Silk screen**   This is sometimes called a Legend or Nomenclature layer. This is the marking that shows the component designations and outlines. You can pick to have either side or both sides silk screened, or with no silk screen at all. This increases the cost somewhat but makes the boards very attractive and easier to assemble. Some vendors will give you a choice of silk-screen colors.

- **Gold fingers**   If your board requires gold-plated edge connectors, you have to specify how many you need, how many edges have these fingers, and the dimensions of each finger.

- **Cutouts**   Sometimes known as slots, this is only required if you have a board that requires one or more slots cut into it. You simply specify how many slots are in the design.

- **Route points**   Rectangular boards have four route points (one in each corner). For an irregular board, each route point is where the router changes direction.

- **Routing**   Most of the time you will specify individual routing—you want separate boards. However, if you want the boards to stay on the larger panel (to facilitate assembly) you can ask for scoring or route/retain. Scoring just cuts partially through the board at each edge. The route/retain uses a router to cut most, but not all, of the board away, leaving the boards connected together by tabs that are easy to break. Some low-cost boards are not routed, by the way, but simply sheared. In any case, some wasted space is left to make way for the cutting tool (so you can't get two 3-inch boards on a 6-inch panel because that would leave no room for routing).

- **Blind or buried vias**   A blind via is one that occurs within the inner layers of a multilayer board. If you are making boards with one or two sides, you will not have any blind vias.

Of course, you may have to enter other things—this list covers most of the common questions, though. Some companies will ask you what lead time you want (in other words, how quickly do you want the boards). Others will show you pricing based on the lead times they suggest (shorter lead times are more expensive).

Quote Sample

Just to show you the kind of difference certain items make on pricing, I used one of the online quote forms from a popular PCB maker. Tables 10-1 and 10-2 show the specifications for a board. Each variation is only slightly different from the baseline board, so you can get a feel for how much the various options change the price. Of course, different vendors may have different pricing structures, but it is still an interesting comparison. Variation A in Table 10-2 would be what I'd consider a standard board.

Table 10-1

Sample board specifications

Layers	2
Material	FR4
Thickness	.062
Copper weight	1 oz.
Size	4x5 (inches)
Number of holes	175
Smallest hole	>.015
Spacing	.008
Solder mask	Both sides
Silkscreen	Top side

Table 10-2

Variations

Variant	Parameters	Value
A	None	
B	Material, layers	Ceramic, 1 layer
C	Copper weight	2 oz.
D	Smallest hole	.010
E	Spacing	.005
F	Solder mask	Bottom only
G	Solder mask	None
H	Silk screen	None
I	Solder mask, silk screen	None, none

If you examine Table 10-3 carefully, you'll see that pricing PCBs must be an art and not a science. You should also remember two things about the prices shown. First, the prices are unit prices. So 100 of the baseline boards cost $377 (in three weeks). Also, a $100 tooling fee is associated with all the two-layer boards (the one-layer board has a $50 tooling fee). If you want electrical testing, the one-time fee goes up even higher.

If you've read this far, you may be wondering if you should go back to Chapter 8 and Chapter 9 make your own boards! The pricing here is from a representative board house in the United States. If you wanted to get boards similar to what you would make yourself, you could look at the I variation. You can get five boards for $220 ($120 for the boards and $100 for tooling). That's a lot—especially if you really only wanted one board. However, a few tricks can drive the price down.

First, you should know that PCB making is a highly competitive business. Nearly all vendors will offer you some sort of discount on your first order. Usually this is enough to offset the tooling costs. One popular board house recently ran a promotion where they would make your first lot of boards absolutely free. They didn't even charge shipping. The downside is that you can only use these discounts once. However, for established customers, most of the board makers will frequently offer specials that can reduce the price.

For example, the same company I used to generate Table 10-2 was running several specials that didn't show up on their online quote form. You could get prototypes (up to 60 square inches) for $25 each (minimum of 3). For 20 square inches and below, the price dropped to $15 each (minimum of 5 pieces). They were not charging tooling costs, so this was actually quite reasonable. To get this price, however, you had to accept a standard set of board parameters. In other words, if you didn't want a .062-inch FR4 board, or you wanted yellow solder mask, or you needed very small holes, you didn't qualify for these prices. So if you could have lived with the prototype board specifications (which are adequate for most boards), the baseline board would have cost $75 (for five pieces). You would have had to add $10 for shipping, but you would have gotten a set of high-quality boards without spending money on chemicals, drills, and other equipment.

Another possibility is to make the board house an offer. Many vendors now have a way that you can set a target price for a board (you have to provide a credit card number, in most cases). If they accept the price, your order is placed and they charge your card. If the board maker is not busy, it makes sense to make some money rather than having expensive machines sitting idle. Small board houses are especially amenable to this.

Table 10-3

Sample Pricing
Table

Variation	Lead time	Unit price (quantity 5)	Unit price (quantity 100)	Percent difference from baseline (quantity 5/ quantity 100)
A (baseline)	3 weeks	$30.00	$3.77	0% / 0%
	1 week	$35.00	$4.82	0% / 0%
	1 day	$70.00	$7.38	0% / 0%
B (ceramic)	3 weeks	$45.00	$4.79	50% / 27%
	1 week	$45.00	$4.98	29% / 33%
	1 day	$70.00	$7.04	0% / −5%
C (heavy copper)	3 weeks	$30.00	$3.95	0% / 5%
	1 week	$35.00	$4.82	0% / 0%
	1 day	$70.00	$7.38	0% / 0%
D (small holes)	3 weeks	$36.60	$4.26	5% / 13%
	1 week	$42.70	$5.43	22% / 13%
	1 day	$85.40	$9.01	22% / 22%
E (tight spacing)	3 weeks	$36.60	$4.35	22% / 15%
	1 week	$42.70	$5.54	22% / 15%
	1 day	$85.40	$9.01	22% / 22%
F (bottom mask)	3 weeks	$28.00	$3.62	−7% / −4%
	1 week	$33.00	$4.67	−6% / −3%
	1 day	$68.00	$7.24	−3% / −2%
G (no solder mask)	3 weeks	$26.00	$3.47	−13% / −8%
	1 week	$31.00	$4.52	−11% / −6%
	1 day	$66.00	$7.09	−6% / −4%
H (no silk screen)	3 weeks	$28.00	$3.62	−7% / −4%
	1 week	$33.00	$4.67	−6% / −3%
	1 day	$68.00	$7.24	−3% / −2%
I (no mask or silk screen)	3 weeks	$24.00	$3.32	−20% / −22%
	1 week	$29.00	$4.38	−17% / −10%
	1 day	$64.00	$6.94	−9% / −6%

They can sometimes run small boards as part of a larger job where material would go to waste anyway.

The other option is to move your board production overseas. This seems daunting at first, but it isn't nearly as hard as you would think, thanks to the Internet and global shipping. For example, Olimex (a Bulgarian company; see Appendix A) will make a 6.3- by 3.9-inch board with roughly the characteristics of the baseline board in Table 10-1 for $26 (with no tooling fees). If your board is small, they will fit as many as possible on the panel for you and cut them apart. I've had boards made there many times and the quality is very good. The downside? Shipping things from Bulgaria (and most of the rest of the world) is fast or cheap, but not both.

Shipping on a small number of boards from Olimex is usually $8. However, the shipment method they use takes about two weeks. This is on top of the time it takes them to process the board. You can have express for $50 (two or three days) or airmail for $30 (about a week). However, at that point, the price difference is not much better than using a domestic board house. When you add $50 for shipping, the panel isn't $26 anymore—it's $76. So you'd get one A variant board for $76 in about five or six days. Using the prototype pricing I mentioned earlier, you could get five boards for $100 in six days (counting shipping costs and time) or for $85 shipped ground, which might take a few days more depending on how far away you are from the board maker.

Importing Boards

If you can afford to wait, the overseas manufacturers are a bargain. I've mentioned Olimex, but you'll find that many different vendors in areas like Central Europe, Asia, and Mexico can provide very inexpensive boards.

I have never had any problems with tariffs or import duties on small quantities of boards. Of course, you should check with the U.S. Customs office (or, if you aren't in the United States, the appropriate authorities) to make sure no import tariffs are due. You can find information about customs at www.trade.gov. Import (and export) goods are classified by a *Harmonized Tariff Schedule* (HST) code.

For boards with no components, the base HST code is 8534.00.00. You can add another two numbers that indicate if the board is fiberglass, ceramic, multilayer, or flexible, but the duty schedule (as I write this) is the same for all of them. Currently, no import duty exists on boards from nearly all countries. Only a few countries have a 35 percent import duty (as I write this, those countries are Cuba, Laos, and North Korea). Of course, you should verify this with U.S. Customs (`www.customs.ustreas.gov/`) because things can change.

The board house you are using probably ships overseas frequently. They can advise you about any special fees and the approximate amount of time required for shipping. I've bought boards from many different countries and it is usually painless. About the only trouble I've run into is that you usually have to sign for an overseas package—they won't simply leave it in your box or at the door.

Paying companies overseas is much simpler now. Although a few companies still want *telegraphic transfers* (TT, usually known as a wire transfer in the United States), most will take credit cards. A few accept online payment services (like PayPal, for example). I prefer to pay with a credit card. That way if you have a problem, you can dispute the charge with your bank. Don't use a debit card unless you are certain you are protected against fraud. Many debit cards don't offer the same protection as credit cards. I have never had any trouble with any of the overseas manufacturers I've used, but it is nice to know you could dispute the charge if you needed to do so. With a company halfway around the world, you may have very little practical legal recourse in the event of fraud.

A TT, or wire transfer, is more complicated and usually adds to the cost since your bank will charge you a fee to send the wire. If you wire money to the wrong place, it would be very difficult (and maybe impossible) to get it back. You also have no protection against fraud as you do with a credit card. I would not use TT unless I had a very large order with a company I trusted. Your bank will give you a form to fill out. They may not do many international transfers, so it pays to have them show you the entire form (after they have filled out their part) before they send it. I've caught more than one mistake this way. You can also do wire transfers online (and it is usually less expensive than using your bank). Citibank offers this service, for example, at www.c2it.com (although not for all countries).

Of course, all of the preceding discussion assumes you are in the United States. If you reside elsewhere, you'll have to find out the relevant laws from your local authorities.

Final Checkout

You should always check your boards carefully before you make them. However, when you are putting up $100 or more for tooling and you are waiting three or four weeks for the finished board, it becomes even more important to make certain that you have the board exactly as you want it. Of course, an electrical rule check and a design rule check are a good start. Some vendors will have Eagle-compatible design rules you can download. Otherwise, you'll need to look at the board house's capabilities and make your own set of design rules.

However, just as a spell checker can't catch poorly written sentences, Eagle's DRC can't catch mistakes that are not technically incorrect, but still result in a board you don't want. Here's a quick checklist:

- Does the board have mounting holes? Are they big enough?
- Are all polarized components (for example, electrolytic capacitors) marked?
- Did you put a copyright notice and part number on the silk screen?
- Are physically large components separated enough to clear?
- Can power transistors and voltage regulators accept a heat sink? Is there room?
- Are all holes large enough? Some items (for example, power connectors) may have thick leads.
- Are all test points and interconnections labeled?
- Is any silk-screen text in awkward places? If so, use the Smash command and move them.
- Do you have a large number of vias? If so, make a copy of the board and try rearranging things to reduce vias.
- What are the two closest features on the board? Are you happy with them? If not, rearrange them and repeat until you are satisfied.
- Are connectors arranged so that plugging them in backwards will be as harmless as possible?

You'll probably add to this list as you make boards. It is very rare that you'll make a board that you are completely satisfied with the first time. If you are lucky, you'll get a board you can live with. Then when you have a new batch made, you can introduce all the corrections you wanted to make. Be sure to keep notes. You'll have a lot of ideas when you first use the board that are hard to remember six months later when it is time to reorder boards.

Another key step is to make a cardboard mock-up board (discussed in Chapter 3). Fit all the components on the board and be certain you are happy with the placement of each component. Imagine you are soldering or servicing the board and try to visualize where components are too close or in awkward positions. It is very useful to have someone else look at the board at this point. It is cheaper to print out another copy of your board and paste it on cardboard than it is to waste a batch of bad boards and have to pay another setup fee.

Of course, if you are making a very large volume of boards, some vendors will make a few samples for you to approve. If you don't like them, they will charge you an additional prototype fee and you can change the layout. When you approve the board, they will then run your large number of boards and essentially you have the correct prototype boards for free.

Using a Gerber Viewer

If your board maker accepts Eagle files, you don't have much to worry about. You do need to agree with the board maker on which layers in Eagle correspond to which layers in the board. For example, if you put custom silk-screen information in a custom layer, you can't expect them to capture that data unless you tell them about it. You may want (or not want) the values to appear in the silk-screen layer—vendors will usually tell you what layers they want, or you can tell them what layers are important to you.

However, if you are using Gerber files, you are acting somewhat on faith. You tell Eagle to make the Gerber files, you send them to the board maker, and you hope the files are correct.

A good board house will contact you if they see something really unusual (for example, a blank layer, or layers that clearly don't line up). However, it is reassuring if you can view the Gerber files yourself.

Luckily, many free tools will allow you to view Gerber files. Some of them are demo versions of larger PCB layout programs that have this capability. By allowing you to view Gerbers for free, the software company hopes you'll check out their product and perhaps buy it. Open-source Gerber viewers are also available (such as gerbv for Linux).

Windows has many choices. You can also download GC-Prevue for free from GraphiCode. This is a limited version of their full-blown software, but it does a good job of viewing many CAM data formats.

Once you install GC-Prevue, you can run it from the Start menu. Just use the File | Import command to add each Gerber or Excellon file to the view. You'll see each layer in a different color and you can easily see if you have any alignment or other problems in the Gerber file.

However, keep in mind that you don't strictly need to view the Gerber file output. However, it does help increase your confidence that the board maker will produce what you have in mind.

In Summary

You get a certain satisfaction from doing something yourself. However, a certain satisfaction also comes from having a nice-looking professional board with solder mask, silk screen, and plated through holes. After all, you've designed a circuit and laid out the printed circuit board, and you still have to build your project. No one would blame you for skipping the messy chemicals and letting a professional shop make your PCBs.

If you have any interest in mass production, you'll find you can't match the efficiency these board makers achieve without considerable expense. Also, if you are working with surface-mount components, you'll find a professionally leveled board with solder mask will make assembly much easier.

The first board you send off is a lot like a first date. You're nervous, you don't know what to expect, and you wonder if you just blew $100. After a while though, it becomes old hat. You just have to learn to be very careful, and don't rush. Every time I have had a board made that was not satisfactory, it was because I was in a hurry to get the board done. Take your time, check, double-check, and check again. It is still less trouble than making your own.

CHAPTER 11

End to End

I've been a ham radio operator since 1977 (my call sign is WD5GNR). When nonham visitors see my radio equipment, they often comment on how complicated it looks. The radio I use currently has 9 knobs and 48 buttons, along with a large *liquid crystal display* (LCD) that has an analog-style meter and several digital displays. At a glance, it seems confusing. However, in practice you only use one or two controls at a time. So what seems like a hodgepodge of controls is actually simple to an experienced operator.

If you've read the first 10 chapters of this book, you've covered a lot of ground. It can certainly seem overwhelming—drawing schematics, creating libraries, routing traces, and performing all the etching steps may seem like a lot to do. It might surprise you to learn that with a little experience, you can take a board from a napkin sketch to a finished board in an afternoon.

Consider the schematic in Figure 11-1. This simple PIC circuit (combined with a program) acts as a logic probe. The three *light-emitting diodes* (LEDs) provide output indicators. The input pin connects to a logic circuit under test. The resistor marked R4 on the schematic allows the PIC to attempt to set the input pin to a different logic state. This allows the PIC to differentiate between a floating input (that is, an input or disconnected wire) and one that some external circuit drives to a logic level. If nothing else is driving the input pin, the PIC will be able to easily drive the input (on pin 9) through R4. However, if the external circuit is driving pin 9, it should have no trouble overriding the weak signal through the 100K resis-

Figure 11-1
A PIC-based logic probe

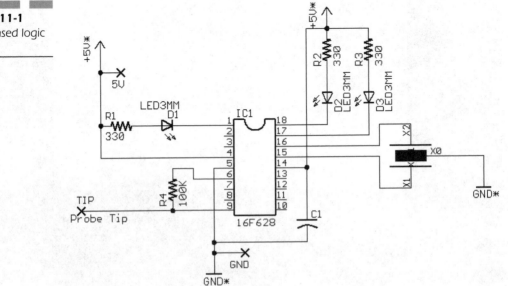

Figure 11-2
The resonator
component

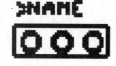

tor. Note that this signal is so weak that even coupling noise from your hand may cause one of the other LEDs to light in conjunction with the open circuit LED. You could reduce this effect by reducing the value of R4 or increasing the delay time in the software. However, you can assume that if the open light is on, the circuit is open (even if another LED is also lit).

A few oddities are present on the schematic. For one, the board has three external connectors: one for the input probe and two for the power and ground connections. The schematic uses a wire pad component to indicate these connections. It seems redundant, but the wire pads for the five volt line and ground are present, as well as the corresponding Eagle supply components. This is necessary so that Eagle understands the significance of the power nets.

I couldn't find a PIC16F628 in the Eagle library on my computer. Rather than draw a custom library, I simply used the standard 18-pin *Dual Inline Package* (DIP) component. This is not quite as handy, since the signal names don't appear on the schematic. However, it is easier than drawing an entirely new library component.

The crystal-like component is a custom library part. This is a standard 3-pin ceramic resonator (see Figure 11-2). The resonator is a 20 MHz unit, so the logic probe operates as fast as the processor will allow.

Board Layout

This board is quite simple, so I decided to fabricate it myself on a single-sided board. When you are making a board by hand, you don't want to make the traces too thin. I settled on 20-mil traces, which shouldn't be a problem on a board this sparse.

I put markings for the three terminals in the bottom-layer copper. This is a good trick for simple markings on a handmade board. Thin marks (made with a | character) also appear near the cathode of each LED. Lettering also helps you to be sure you aren't accidentally using mirror image artwork.

Figure 11-3
The routed board

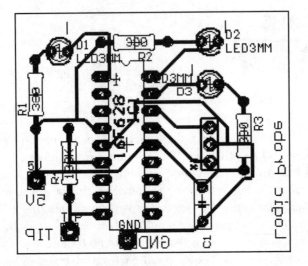

Since this board was handmade, I didn't mark mounting holes on the board. Instead, I etched the board on an oversized piece of laminate. Then it was a simple matter to hand-drill mounting holes to fit the board into the case. There was no reason to automate the mounting holes in this case.

In all, this is a fairly typical homemade board. I manually routed the resonator leads because these are critical and you want the traces to be direct. After that, I let the autorouter lay out a single-sided board (see Figure 11-3) —it did a satisfactory job.

Production

The next step is to print the board layout. Of course, you will turn off all the layers except the bottom and the pads before printing. The board does not have vias, and none of the other layers affect the bottom layer of copper. I used the Print dialog to select black and solid output. All other options were off. Since Eagle prints as it looks down through the board, you do not need to mirror the bottom layer (since it is already mirrored).

The printer I used was an HP Laser Jet 4000 and the paper was Epson matte finish photo-quality inkjet paper (Epson #S041062). An iron on the permanent press setting transferred the toner to a small piece of double-sided copper laminate. The pattern appears to bleed through the back of the paper when the pattern transfer is complete. I didn't have any single-sided boards, but it is simple enough to etch away all the copper from the top side (although it is a bit wasteful of your etchant).

Wetting the Epson paper allows you to peel it away from the board easily. However, a thin layer of paper remains over the entire pattern. A quick rub with a brush (like a toothbrush or a dishwashing brush) will remove this paper film, leaving a good pattern (see Figure 11-4). Some of the larger pads have paper fibers embedded in the toner, but this is not a problem. You also have to be very careful to get the paper out of the interior parts of the lettering.

At this point, you should inspect the board to make sure the pattern is clear. If the resist has run together, you may be able to salvage your work by using a hobby knife. If the resist has folds, a permanent marker can save the day. In this case, the resist transfer was essentially perfect. The only thing that appeared too thin were the marks near each LED, so I just dabbed them with a marker to make them heavier. This is a good time to notice the narrowest gaps in the resist pattern. When you etch, you need to make sure all the copper is gone from these areas (otherwise, you'll need the hobby knife again).

I used a RubberMaid tank with ferric chloride for the etching. An aquarium air pump supplied agitation, and I used an aquarium heater to heat the etchant. The board took about 15 minutes to etch (it would have been quicker if I had preheated the etch). The resulting board appears in Figure 11-5.

The next step is to drill the holes in the board. I used a small carbide drill bit in a Dremel tool with the drill press attachment. I simply eyeballed most of the holes (magnifying binoculars help). However, for the *integrated*

Figure 11-4
The board ready
for etching

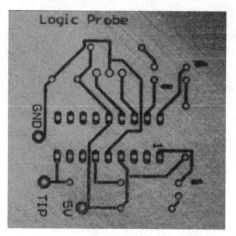

Figure 11-5
The board after
etching, but
before drilling

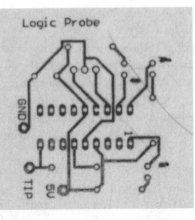

Figure 11-6
The logic probe
board after
assembly

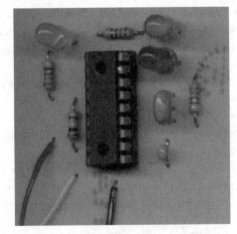

circuit (IC), I used masking tape to secure a piece of perforated board aligned over the holes. This makes it very easy to line up the drill bit precisely, turn on the drill press, and make the hole. After I drilled the holes, I put the board in a shallow bath of Liquid Tin for about two minutes. This produces a good-looking protective tin coat.

Assembly

With the finished board, it is simple to assemble the board (see Figure 11-6). I used an 18-pin socket because I wanted to be able to take the chip out for reprogramming. The tin plate improves the solderability of the board. However, the board has no solder mask, so you have to be careful not to bridge adjacent traces.

In Use

The circuit by itself does not do anything. The PIC microcontroller requires a program before it will actually do anything. Luckily, programmers for this type of PIC are readily available (and many low-cost or free plans are available on the Web). You also need the free software (MPLAB®) from Microchip if you want to change the code.

The CD-ROM contains the hex file required to program the PIC (consult your programmer's documentation to find out how to transfer this hex file to the chip). If you want to change the program (see Listing 11-1), you can open the .PJT file (from the CD-ROM) using MPLAB, modify the .ASM file, and then recompile the hex file.

Listing 11-1

The logic probe software

```
;*********************************************************************
; Logic Probe                                                       *
;*********************************************************************
;*    Notes:                                                        *
;*        This program determines and displays                      *
;*        the state of an input                                     *
;*        pin. It can detect a zero, a one,                         *
;*        or open state.                                            *
;*********************************************************************

list      p=16f628      ; define processor
#include <p16f628.inc>            ; processor definitions
; Note: the line below splits in print, but should be
; one single line (starts with _CONFIG and ends with
; _LVP_OFF)
        __CONFIG _CP_OFF & _WDT_OFF & _BODEN_ON & _PWRTE_ON & _HS_OSC &
_MCLRE_ON & _LVP_OFF

;***** VARIABLE DEFINITIONS
w_temp          EQU       0x70        ; ISR context saving
status_temp     EQU       0x71        ; ISR context saving

; I/O definitions
#define drive PORTB,0
#define tip   PORTB,3
#define ledlow PORTA,0
```

(continued)

Listing 11-1
(continued)

The logic probe
software

```
#define ledhigh PORTA,1
#define ledopen PORTA,2




;*********************************************************************
        ORG     0x000               ; reset vector
        goto    main                ; go to start

; I am not using interrupts, but I left the standard
; handler here for future expansion
        ORG     0x004               ; interrupt vector
        movwf   w_temp              ; save context
        movf    STATUS,w
        movwf   status_temp

; isr code can go here or in a subroutine
        movf    status_temp,w       ; restore context
        movwf   STATUS
        swapf   w_temp,f
        swapf   w_temp,w
        retfie                      ; return from interrupt



main:
        ; turn off analog pins
        movlw 7
        movwf CMCON
        ; initialize (turn unused pins to output)
        movlw 0xFF
        movwf PORTA     ; set all LEDs off
        bsf STATUS,RP0
        clrf TRISA      ; all PORTA to output
        movlw 0xFE      ; only the drive is an output
        movwf TRISB
        bcf STATUS,RP0

loop:   bcf drive       ; try to force tip to 0
        goto $+1        ; wait for stable state
        goto $+1
        btfsc tip
        goto sethigh    ; if it is still high, then it is really high
        bsf drive       ; now try to force tip to 1
        goto $+1        ; wait for stable state
        goto $+1
        btfss tip
        goto setlow     ; if it s 0, then it is really 0

; if we get here then the pin is either tristate or pulsing
; faster than we are running
        bcf ledopen
        bsf ledhigh
        bsf ledlow
        goto loop
```

Listing 11-1
(continued)

The logic probe
software

```
sethigh:
        bcf ledhigh
        bsf ledlow
        bsf ledopen
        goto loop
setlow:
        bcf ledlow
        bsf ledhigh
        bsf ledopen
        goto loop


        END
```

The program is actually quite straightforward. The section between the main and loop labels initializes the PIC. It is important to set the input and output pins, and also important to turn off the PIC's analog comparator function so that the shared analog pins can drive the LEDs.

The main loop tries to force the input to a particular state, delays, and then samples the input. If the input is in the opposite state, then some other circuit is driving the input and the program lights the appropriate LED. If the PIC can successfully force the input's state, then it assumes the input is not connected to anything and lights the open-circuit LED.

In Summary

The project in this chapter is an excellent first step to designing and fabricating your own boards. Working with a simple design allows you to perfect your technique and assemble your equipment. Once you have the process nailed down, you'll find it is relatively simple to produce boards, even if they are more complex than this one.

Another similar (and simple) project appears in Chapter 4, "Autorouting." This project also uses a PIC (albeit, an 8-pin PIC). Once you've tackled a few simple projects, you'll be ready to produce larger boards of your own design.

CHAPTER 12

More Choices

I've often wondered if we had eight fingers on each hand would we count in hexadecimal instead of decimal. There is no denying that your tools influence you in subtle but concrete ways. Pictographic languages with many elaborate symbols (such as those found in many parts of Asia) are much simpler for handwriting recognition than simple sets of glyphs such as English. On the other hand, those pictograms make it very difficult to develop practical keyboards.

I've heard it said that when all you have is a hammer, every problem looks like a nail. That's probably true with most computer tasks as well. If you are an accomplished C programmer, you probably want to write programs in C. If you know how to use Microsoft Word, that knowledge influences how you create documents.

This book covers one specific tool: Eagle. You can probably guess that it isn't the only program around. I selected Eagle for this book for several reasons:

- A very useful version is available for free.
- It is relatively inexpensive to upgrade the software.
- The program has sophisticated schematic capture and autorouting capabilities.
- The Eagle library is extensive, and you can add to it, even in the free version.
- You can run Eagle on Windows or Linux.

Those are some compelling arguments. For personal use, you can use the free version of Eagle and still enjoy high-end features like schematic capture and autorouting. Many free programs only demonstrate the application and provide skimpy libraries, prevent you from creating your own libraries, or don't include advanced modules. Others limit you to a certain amount of time that you can use the program without charge. Eagle has, wisely in my opinion, elected to let you use their full program with only size constraints in the free version.

Of course, Eagle isn't perfect. In particular, its user interface is a bit different from a "normal" Windows program. It occasionally crashes, and if your board and schematic get out of sync (perhaps from a crash), you could be in serious trouble. Navigating the vast libraries isn't always easy (although at least the library is, in fact, vast). However, nothing is wrong with Eagle that you won't get used to with some practice.

Other Contenders

Many *printed circuit board* (PCB) programs are available. Here are a few others that offer some sort of free demo (or, in some cases, offer a full version for free):

- **AutoEngineer** Bartel's supplies this integrated package that supports their well-known autorouting package.

- **CirCad** This Windows package from Holophase claims to be the fastest performing *computer-aided design* (CAD) tool. One unique feature: You can scan an existing board and generate a schematic.

- **Circuit Creator** AMS, the creator of this program, offers a 14-day trial.

- **Circuit Maker 2000** This is a complete suite of programs that includes schematic captures, simulations, and PCB layouts.

- **Douglas Electronics** Douglas produces a native Mac CAD program (with a Windows version also available). The low-cost version limits you to their production facility or printing layouts on your own printer. For a steep additional price, you can buy the same software with Gerber output.

- **EdWin** EdWin is from Swift Designs in Britain and you can download a trial.

- **Electronics Workbench** The Utilboard module of this suite (which includes simulation) can lay out boards with up to 32 layers with a 50-inch by 50-inch size.

- **PCB** An open source layout tool with an extensive library. However, you have to be somewhat of a Unix hacker to extend the program (and, in some cases, install it).

- **Proteus** This British board layout program, like Eagle, is available in several levels.

- **MUCS PCB** This is free software from the University of Manchester. It uses a text-based entry and so is not as friendly as Eagle or most other programs in this list.

- **SuperPCB** Another Windows-based suite (including simulation) from Mentala.

- **WinBoard**   From Ivex, this is another Windows-based program with a related schematic capture module.
- **WinPCB**   This is part of another Windows suite from Interface Technologies. The fact that you have to mail them for the price means it is probably more than you want to spend!
- **XCircuit**   This program is an open source editor that can draw many things, including PC boards.

Many other vendors offer programs: Altium, AutoCAD, Cadence, and Mentor Graphics are just a few. Many of these are high-end packages (some do more than just PCB layout) and are typically out of reach financially for most small businesses and hobby users. Some of these have older versions or student versions available. Most notably, Protel (now owned by Altium) makes DOS versions of its EasyTrax and AutoTrax programs available for free. If you can stand working in DOS, these programs were high-end packages in their day. You can also get time-limited trial versions of many of these programs. However, be aware that some of them cost quite a bit of money. Some professional packages sell for well over $10,000 with all the options and features included. As I write this, Mentor Graphics's PADS, for example, is on sale for *only* $8500!

Most of the open-source tools are made for Linux. However, if you are adept at compiling Linux programs, you may have some success porting these programs to Windows using the excellent (and free) Cygwin library (see http://sources.redhat.com/cygwin). In many cases, someone has already done this, and there will be a Windows version of the tool available for download.

Companies that make CAD software seem to buy each other out on a regular basis. Sometimes web sites change, so if you are looking for links to any of the software mentioned above, look in Appendix A for the most recent URLs.

Things to Look For

If you are evaluating other packages, you should look at a few aspects of the programs:

- The program should be able to output to Gerber files and print to your output device. Avoid programs that only write to proprietary formats or force you to use a particular vendor to make boards.

■ No matter how good the included library is, you'll need custom components sooner or later. Make sure you can easily add libraries.

■ If the package has an autorouter, be aware that autorouters vary wildly in their effectiveness. The Eagle router is a rip up/retry router and performs very well. Some autorouters use a grid router, which is often less effective.

In Summary

Like most things in life, people tend to polarize on their choices. Many people have their favorite tools, and they are certain you should use them instead of other possible selections. On the one hand, it is good to try a few tools and see what you like. The availability of free trial versions makes this very convenient. On the other hand, fully learning to use a PCB program can be a big investment in time, so you probably don't want to learn dozens of different programs.

The open-source programs are certainly tempting because of the price. However, I personally haven't seen any of the free programs that are as capable as Eagle using my criteria (Gerber output, ripup router, and powerful library). If you need larger boards than the basic Eagle program provides, and you are on a budget, you might want to try some of these free programs. However, you'll find that you often get what you pay for—so if you really need more capability, you may find that Eagle is still a bargain.

Of the commercial offerings, only you can decide which programs you like. Many capable programs are out there. Some cost more than Eagle and some cost less. Ultimately, though, a program that doesn't work well or won't let you design a board isn't a bargain at any price.

APPENDIX

Resources

There are many board makers, supply companies, and software companies that are useful to people designing or producing PC boards. Here are just a few that I've encountered. Of course, this list could never be complete. In addition, companies change names, are sold, or simply shut down from time to time. A Web search should turn up many other options. I don't have personal experience with all of these companies, so be sure they will meet your needs before placing an order.

Board Makers

Accutrace—http://www.pcb4u.com

Advanced Circuits—http://www.4pcb.com

AP Circuits—http://www.apcircuits.com

CustomPCB—http://www.custompcb.com

E-Teknet—http://www.e-teknet.com

Imagineering—http://www.pcbnet.com

Olimex—http://run.to/pcb

PCB Express—http://pcbexpress.com

PCB Fab Express—http://www.pcbfabexpress.com

NOTE: Searching Yahoo's B2B Electronics directory for PCB will yield many more companies.

PCB Software (commercial)

Vendor	Product	URL
Altium	Protel, P-CAD, Circuit Maker 2000	http://www.altium.com/
AMS	Circuit Creator	http://circuitcreator.com
AutoDesk	AutoCAD	http://www.autodesk.com
Bartel	AutoEngineer	http://www.bartels.de
Cadence	Allegro, OrCad	http://www.cadencepcb.com
CadSoft softusa.com	Eagle	http://www.cadsoft.de, http://www.cad-
Douglas Electronics	Douglas CAD/CAM	http://www.douglas.com
Electronics Workbench	Utilboard	http://www.electronicsworkbench.com
Holophase	CirCad	http://www.holophase.com
Interface Technologies	WinPCB	http://www.i-t.com
Ivex	WinBoard	http://www.ivex.com
Labcenter Electronics	Proteus	http://www.labcenter.co.uk
Mentala	SuperPCB	http://www.mentala.com
Mentor Graphics	PADS, Expedition, Board Station	http://www.mentor.com

Open Source or Free PCB Software

AutoTrax EDA—http://www.autotraxeda.com/

PCB—http://bach.ece.jhu.edu/~haceaton/pcb

MUCS PCB—http://www.cs.man.ac.uk/apt/tools/mucs_pcb/

XCircuit—http://bach.ece.jhu.edu/~tim/programs/xcircuit/

PC Board Supplies

Circuit Specialists—http://www.web-tronics.com

Injectorall—http://www.injectorall.com

MG Chemicals—http://www.mgchemicals.com

Radio Shack—http://www.radioshack.com

Tools, Supplies, and Surplus

Alltronics (http://www.altronics.com)—Often has surplus drill bits and other items.

Drill Bit City (http://www.store.yahoo.com/drillcity)—Drill bits.

Electronics Goldmine (http://www.goldmine-elec.com)—Often has surplus bits and laminate.

Harbor Freight (http://www.harborfreight.com)—General tools, carbide drill bits, gloves, masks.

MECI (http://www.meci.com)—Surplus company; often has laminate.

MicroMark (http://www.micromark.com)—Small tools, drill bits, precision drill presses, indexing tables.

Related Software

Artwork Conversion (www.artwork.com)—Produces XGBRVU which allows you to view Gerber files.

gerbv (gerbv.sourceforge.net)—This is an open source Gerber viewer for Linux.

GraphiCode (www.graphicode.com)—You can use the free GC-Preview program to view Gerber files.

PentaLogix (www.pentalogix.com)—The demo version of the ViewMate program can view Gerber files (formerly Lavenir).

INDEX

S

T

ABOUT THE AUTHOR

Al Williams is the author of 18 books, including many programming titles, as well as *Embedded Internet Design*, also published by McGraw-Hill. An electronics professional for over 20 years, he is principal of his own company, AWC, which produces hardware modules for microcontroller developers. Mr. Williams also designs PCBs for a wide array of clients.